THE SEVEN-YEAR FIX
A comprehensive global solution

Jan Moxley

ZONE INTERACTIVE
MOUNT DORA, FLORIDA

Copyright © 2016, 2019 by Jan Moxley   All Rights Reserved

Library of Congress Control Number: 2016910391

ISBN: 978-1-887926-01-0

This book is dedicated to my twins, Kayla and Jessica, my sunshine, soulmates, and loves of my life.

Apono Kayla Blue   *Jessica Kai*

Special thanks to Jean-Michel Cousteau and Dr. Patch Adams for their grand contribution and to Heather Rogers, author of *Green Gone Wrong*, which inspired me to write this book.

CONTENTS

PREFACE

Since 2008 my full-time goal has been to propose a solution to global warming, food and water shortages, electric power sustainability, and out-of-control healthcare costs. This is the primary objective with special emphasis on the elimination of famine and poverty and protection of wildlife (those without a voice). A comprehensive solution had eluded me for years, but recent technological developments, and worldwide sentiment and demand, has finally advanced to the point of public acceptance of a global initiative. Governments have failed miserably. Now is the time to stipulate resolve and promote immediate action.

As the point of no return rapidly approaches, the people must take the lead and the leaders must follow. Those that believe a vast global solution is too expensive, I ask this question:

What value do you place on saving our planet?

1
PROJECT CONCEPT AND DESIGN

And there I was… thinking that I must be out of my mind to propose an idea of such an enormous undertaking. A Ranch Farm Cooperative green community or, even more preposterous, 1,166 RFC green communities linked by a magnetic levitation (Maglev) transit distribution network. Designed to function as a global system to supply the world's population with food, water, power, and healthcare for a low yearly membership fee. A per-person membership fee that is based entirely on economics and currency valuation of the country where a member resides, broken down into twelve monthly payments.

When I considered my dislike for co-ops, which is an attitude that had developed over several decades of observing poorly run co-ops, the very thought made me question my overall sanity. There are some co-ops that have worked, however, and those are the ones that make this idea seem not so farfetched. My theory on why co-ops generally fail, besides being poorly run, is that they aren't cooperating on a large enough scale.

So the conception of the Ranch Farm Cooperative (RFC) green community was born; an immense 1,950-square-mile community that produces enough potable water, organic agriculture and livestock, and aquaculture to feed over 6,000,000 people. Owned by 125,416 local area workers, each RFC will be constructed in a designated free-trade zone, and will provide numerous jobs in poverty-stricken, developing, and rich industrialized nations. Every RFC membership will include renewable-generated electricity, produced without greenhouse gas emissions, and a complete healthcare package that includes all medical, dental, and mental health services regardless of age. Hospitals and medical centers will be located within each RFC green community and outside the community in regional locations for convenient access by members.

What I am proposing is a comprehensive global solution to the substantial problems the world is currently experiencing. Governments planned out a resolution for global warming in a 2015 climate conference in Paris. I was skeptical prior to the conference. Now I am dismayed that the best plan they could agree on doesn't begin until 2020. The accord opened for signatures at the U.N. on Earth Day, 22 April 2016. The treaty is now binding after more than the required 55 countries, accounting for more than 55% of the world's greenhouse gas emissions, ratified it. This is an epic achievement, but illusory. The only real accomplishment is that they have managed to ratify an agreement that is both delusional and inadequate, considering it a bridge between current policies and climate-neutrality before the end of the century. [1]

I remember the climate conference in Copenhagen in 2009, and the three words that seemed to resonate throughout the conference: "…without economic hardship…." The very fact that the governments of the world even considered not doing anything because of economic hardship, if they posed too many restrictions on the very entities that are emitting the pollutants, rather than resolving the global warming crisis, was entirely irresponsible and more disappointing than the continuing denial from climate change skeptics. At the 2015 Paris conference, our leaders were congratulating themselves for agreeing on an inept climate plan that fails to achieve any meaningful solution as the clock ticks down to imminent annihilation. Proving once again that governments of the world have no idea how to accomplish the enormous task of saving our planet from an impending environmental catastrophe. [2][94][95][135] [159]

Climate change, however, isn't the only problem we are currently facing. And the pace that governments are moving to resolve global warming, and these other significant global issues, is akin to the story of the tortoise and the hare, but without a happy ending. Slow and steady wins the race is not going to be the moral of this story. The problems (Hare) are just too fast for governments (Tortoise) to address. [136]

It has been apparent for many years that the solution to these global problems has eluded governments for so long that all they are capable of doing, along with the thousands of nonprofit organizations, foundations, and NGOs, is either treat the problem with a bandage or, even worse, promote awareness of the problem in lieu of a plan to actually solve the problem. The Paris accord is not a plan. It is an insult. After forty years I am overly aware and completely perplexed that no immediate solution has been implemented on a global scale when the future of our fragile planet, and all that live on it, is at stake.

Governments were never intended to resolve the world's massive environmental problems, but somehow over the years the responsibilities have been placed on governments that are not equipped to tackle these out of their jurisdiction issues. However, the citizens of each country were instrumental in selecting the unqualified leaders that have failed to resolve the challenges associated with the tasks of implementing solutions to these global problems. Then they fervently complain about them being in office, much to the chagrin of the constituents that elected them in the first place. In truth, every country has the government it deserves.

Besides shelter, there are four essential requirements for modern human life on this planet: food, water, power, and healthcare. However, these essential requirements have become four global problems needing immediate attention. The growing scarcity of food and water, the lack of reliable, environmentally safe electric power generation, and the out-of-control costs of healthcare are every bit as important as climate change. [167]

Food should not be commodities traded on the financial markets. Water should not be a commodity so scarce that the rights are sold off to the highest bidder. Power companies should not be allowed to inflate electricity rates to consumers to pay exorbitant salaries to executives and increase the value for shareholders. Healthcare should not be a business designed to make a profit on people's ailments and injuries. These are the essential requirements of human life and should be provided to every man, woman, and child for a reasonable fee that can be afforded by all people in one's own country and should not be manipulated to make a quick profit.

Before the capitalists start yelling socialist or communist, let me be clear: I am very much a capitalist. Throughout history, capitalism has been the greatest motivator for innovation. I am not against capitalism by any means. What I am against is greed and placing massive profits ahead of our civilization's survival.

The big question is how do we pay for this enormous expense to initiate this plan for the benefit of the world? And even bigger question: How do we pay back the money used to build this vast network of RFC green communities, Maglev transit, distribution terminals, ocean water desalination plants, hospitals, and electric micro-grids?

Unlike the elected leaders in countries throughout the world that continue to implement programs that their country can't afford and have absolutely no inclination of how to pay for them, which over time has brought numerous countries closer to bankruptcy without any hope of ever repaying incurred debts of increasing magnitude. This RFC plan had to have a realistic solution to both funding and payback without any government financing that would plunge countries further into debt.

Currently, the United States has an external national debt of over $22 trillion plus over $3 trillion for state and local external debts, over $6 trillion in unfunded pension liability, $62 trillion for their Social Security fund, after raiding the coffers for decades, and another $5 trillion for the government-backed mortgage bailout. Additionally, total personal debts of all U.S. citizens are over $19 trillion. The estimated total of the external national debts of all other countries is over $60 trillion. With a combined global debt of over $177 trillion the likelihood of a payoff is nowhere in the near future. But even more disheartening, the likelihood of a solution to the global warming crisis, especially after the 2018 climate conference debacle in Katowice, is not in the near future either. [3][4]

Which brings us back to the question: How do we pay for this immense global project without government backing? When developing the RFC green community plan several potential problems had to be considered. The first was how to keep the RFC owners in control without any outside influence. This is a concern that private and publicly traded corporations face on a reoccurring basis. The problem is shareholders that can use their collective percentage of stock ownership in a company to manipulate the hiring and firing of presidents, CEOs, board of directors, etc. Using their might to actually, in some cases, change the direction of a company's business plan in order to raise the stock's value. This was the first problem that needed to be addressed. After careful consideration, I came to the conclusion that a high-yield twenty-year bond was the only way to keep this situation from occurring.

By issuing bonds to fund this global project the trillions of dollars needed to implement this RFC green community and global distribution plan can be raised. If you think that this goal can't possibly be achieved, I offer this fact: There are over 600,000 nonprofit organizations bringing in over $2 trillion per year in donations. Philanthropists, concerned and well-meaning patrons, make their donations to their preferred charities

4

without ever thinking of how their donations are spent. Only that their conscience is clear about giving to a perceived good cause while reaping a reward of income tax deduction. Fair enough, but a large percentage of those donations go to administration expenses and advertising for more donations. In many cases these organizations are competing with other similar organizations and are duplicating their efforts, which negates their overall effectiveness.

No place was this ever more evident than in Haiti after the 12 January 2010 earthquake, where there were a thousand organizations on the ground covering the same small area with absolutely no coordination or cooperation between the groups. And that is the overall problem. I believe that these nonprofit organizations, foundations, and NGOs have a useful purpose in this RFC green community plan and their intended goals, if coordinated together, can be much more efficient and effective than they are separately.

Would there still be over $2 trillion per year in donations if the tax deductions weren't in place? Unknown, but my guess would be no. Debating this rhetorical question though is not a prime concern because donations will not work for a project of this immense size and expense. The cost is much too great. With that in mind, if philanthropists received tax deductions for investments in bonds, then those donors would likely purchase the bonds instead of making a less significant tax-deductible donation. Of course world governments must agree to treat these RFC bond purchases as tax-deductible purchases, but that should not be a concern if governments are serious about fixing the problems at hand. However, this is only one aspect that all countries must agree on to move forward with this RFC global solution.

The Paris conference did accomplish one objective: It compelled every nation to agree a climate problem exists. An epic achievement, but far from a solution. In our predicament solving the problem too late isn't a sufficient remedy. This comprehensive global plan provides a solution today, but governments must ratify these requirements to guarantee its success:

1) Designate every RFC green community and distribution terminal a free-trade zone with unconditional tax autonomy.

2) Grant every RFC owner-worker, and working resident, income tax-free status for RFC-related income within each RFC community.

3) Grant every RFC community a sales tax exemption for purchases of supplies and manufactured goods. Anything purchased for the RFC community is exempt.

4) Grant sales tax exemption status to every RFC member for annual membership dues and for any extra RFC-related sales that may occur outside what is covered in the membership program. For example, unnecessary elective cosmetic surgery, excessive electricity or water consumption, and any purchase of RFC-manufactured goods that are extra charges billed to a member by their RFC.

5) Grant tax-free status on RFC members' electricity and water usage, and on road taxes for RFC delivery trucks, and trucks contracted for delivery, outside RFC communities and Maglev transit distribution terminals.

6) Grant each RFC bond purchaser a tax deduction for the purchase of bonds and an income tax-free classification on each bond's twenty-year maturity payout.

7) Grant RFC communities governing jurisdiction and legal autonomy within their RFC boundaries.

8) Within each country's borders: RFCs pay for their land up to $3,000 per acre. The country then pays all of the costs over $3,000 per acre for RFC green communities, Maglev tracks, distribution terminals, electric micro-grid sites, and water desalination plants. And grants a right of way easement through any non-RFC land that is required for the Maglev transit network, water or fuel pipeline, or any distribution conveyors located between RFC community lands.

9) Each country is required to place easements in a land trust for each RFC community. Then after a preset time limit, maximum 99 years, grants ownership of the land easements to the RFC.

10) Grant flythrough airspace rights and docking rights, tax free, to all RFC community-owned hydrogen airships.

If the countries of the world agree to these ten requirements, then all they will need to do is ratify a mandatory requirement for non-RFC members, which are located outside the RFC community boundaries, to either supply their own, or be supplied, non-greenhouse gas-emitting electricity generation. Either way, the requirement must be unanimous between all countries and completed in the same seven-year timeline as the RFC green community plan. Then governments will have fulfilled their obligation and over the next seven years these global issues will all be resolved. There are fifteen advantages to this RFC global solution:

1) Workers own the RFC green communities that will be located in the designated free-trade zones.
2) Each RFC's 24-hour ordering and delivery system will be servicing a minimum of 6,000,000 members.
3) RFC communities eliminate government subsidies that are currently paid to farms and ranches.
4) Existing farms, ranches, and aquaculture operations that fit into the RFC model will be invited to merge their land and operation into a RFC green community being constructed in their region.
5) No drought problems, because each RFC will produce its own water.
6) No greenhouse gases will be emitted from RFC communities, power micro-grid sites, trucks, or the Maglev transit network.
7) Any possible type of food contamination can be located, isolated, and contained quickly by computerized ordering and delivery tracking, which will lower the chances of any possible epidemic.
8) Annually, members select the types of food produced in their RFC.
9) Major disasters will be responded to quickly by RFC airships. Each RFC will manufacture and store for emergencies: food, water, power, blankets, and temporary shelters, to be ready for immediate loading and transport. All RFCs will be equipped with airship transportable emergency medical facilities and emergency medical airlift drones.
10) RFC communities' owner-workers and members will have complete healthcare access and medical services provided to them under their RFC memberships. Healthcare includes medical, dental, and mental health services.
11) Simultaneous construction of 1,166 RFC green communities globally will create 146,235,056 full-time RFC owner-worker jobs and will create additional non-RFC related jobs within each RFC community.
12) Construction of the Maglev transit network will employ a labor force of approximately 29,790,800 workers.

13) The additional number of jobs that will be created in companies that build machinery, equipment, and the related components necessary to construct and supply each of the RFC green communities, Maglev transit network, trucks, ocean water desalination plants, distribution terminals, electric micro-grids, etc., will boost the global economy.
14) The increased tax revenue from the improved global economy will allow countries to continue to pay their financial obligations without going into default.
15) Purchasers of RFC twenty-year bonds will receive a tax deduction on their RFC donation-investment, and a 4% compounded yield income tax-free return on their bond maturity payouts.

The membership dues for each RFC green community are for the operational expenses listed below. The top membership dues of $4,200 per year USD (€4,200 in the E.U.) are at the top of the scale. Membership dues and expenses listed under Developing Countries and Poorest Countries are for example only. Every RFC will have its membership dues, owner-worker salaries, and the other expenses adjusted for the currency valuation and the economics of the country where the RFC is located, and the countries where its members are located. Each annual membership fee will be divided into twelve monthly payments.

Rich Industrialized Countries

$4,200 – RFC yearly membership: food, water, electricity, and healthcare.

$25,200,000,000 = Minimum 6,000,000 members x $4,200 per year.

Use of yearly revenue

$ 3,200,000,000 – RFC operating and maintenance expenses per year.
  1,000,000,000 – Expansion and future equipment replacement fund.
  1,800,000,000 – Cell phone and internet communication service.
  9,029,952,000 – 125,416 RFC salaries at $72,000 each.
  7,200,000,000 – Complete healthcare for RFC owners and members.
  1,800,000,000 – 40,500 salaries. RFC hospital staff for five hospitals.
  1,170,048,000 – Part-time worker payroll and credits toward college education.
$25,200,000,000 – Total operating expenses per year.

Developing Countries

$1,836 – RFC yearly membership: food, water, electricity, and healthcare.

$11,016,000,000 = Minimum 6,000,000 members x $1,836 per year.

Use of yearly revenue

$ 1,514,408,960 – RFC operating and maintenance expenses per year.
     1,000,000,000 – Expansion and future equipment replacement fund.
        792,000,000 – Cell phone and internet communication service.
     3,943,079,040 – 125,416 RFC salaries at $31,440 each.
     1,800,000,000 – Complete healthcare for RFC owners and members.
        792,000,000 – 40,500 salaries. RFC hospital staff for five hospitals.
     1,174,512,000 – Part-time worker payroll and credits toward college education.
$11,016,000,000 – Total operating expenses per year.

Poorest Countries

$330 – RFC yearly membership: food, water, electricity, and healthcare.

$1,980,000,000 = Minimum 6,000,000 members x $330 per year.

Use of yearly revenue

$ 107,000,000 – RFC operating and maintenance expenses per year.
     1,000,000,000 – Expansion and future equipment replacement fund.
        144,000,000 – Cell phone and internet communication service.
        300,998,400 – 125,416 RFC yearly salaries at $2,400 each.
        240,000,000 – Complete healthcare for RFC owners and members.
        144,000,000 – 40,500 salaries. RFC hospital staff for five hospitals.
         44,001,600 – Part-time worker payroll and credits toward college education.
$ 1,980,000,000 – Total operating expenses per year.

The RFC green communities' expansion and future equipment replacement fund must stay consistent. And although the operating and maintenance expenses and the communication and healthcare budgets per year per country will vary, the level of these services in each country will be equal to the same high-quality standards as they are in any rich industrialized nation.

To eliminate famine and extreme poverty in the world's poorest countries, the Global Survival Fund's online donation program will be set up for rich industrial and developing nations' citizens to donate yearly RFC memberships to those individuals and families that are destitute. See Section 16: Conglomerate.

RFC bond payoff funding cannot come from membership dues. It will come from two other sources. The first is Maglev transit passengers and commercial freight. Operating at extremely high speeds the Maglev flatcars will operate 24 hours a day, every day. The commercial shipping containers and passenger transport container compartments will have a global track capacity of 22,500,000 flatcars per day in addition to the delivery containers shipped from each RFC overnight to their members' regions.

The second source of bond payoff funding is high-volume online advertising. RFC members will order food and water produced by their RFC 24 hours in advance for next-day delivery. Orders will be placed using the RFC website or cell phone app. Every time a member orders from their RFC an advertisement video or banner displays on their app's ordering screen. Each displayed advertisement fits the exact demographic of the particular member placing the order. Billions of RFC members will place orders daily.

All RFC green communities and the Maglev transit network will be in full operation at the end of year seven. The first bond maturity dates will be in year twenty and the last bond maturity payments will be in year twenty-seven. If this global project begins in 2020 (the same year the Paris Agreement begins), then the RFCs would be completed and operational in 2027. And in 2047, all of the outstanding bonds would be completely paid off.

In 2027, the world's goal of eliminating fossil-fueled power plants would be successful and fossil-fueled vehicles and machinery would be completely phased out. The solutions to food and water shortages would be in effect and ameliorate healthcare. See Section 13: Global Funding, and Section 21: RFC Forty-Year Timeline.

Not everyone will be thrilled with the consequences of this RFC green community and Maglev transit plan. During the seven years it will take to implement this global solution, several types of businesses will either go out of business or radically change their business plans in order to stay in business.

The companies that will be the most affected will be genetically modified seed companies, pesticide companies, corporate-owned farms and ranches, fishing conglomerates, food-related processing companies, grocery store chains and their distributors, health insurance companies, hospitals and medical centers, shipping companies, airlines and railroads, oil and gas companies and their suppliers, coal companies, and electric power companies. The power grid, as we know it, will be eliminated.

The prediction of global economic hardship, however, won't be the big issue that was claimed in Copenhagen in 2009. The number of new jobs created by this plan will be plentiful and although many people will feel somewhat displaced by their current jobs being terminated, similar jobs for their experience will be available.

2
LAND REQUIREMENTS

The RFC green communities will eliminate famine and extreme poverty by developing thousands of small organic and regenerative farms and ranches to form the cooperative. Using their new innovative methods and processing techniques will increase crop yields and decrease distribution expenses, thereby lowering total food cost.

Many determining factors must be seriously considered in their planning. Every RFC must be constructed with wildlife habitat protection in mind. Each will be designed to keep natural wildlife migration routes free and protect wildlife refuges surrounding each RFC green community from illegal hunters, trappers, and poachers. Unlike numerous regions throughout the world that allow business and residential developments to continue to encroach on wildlife habitats without thinking about the consequences, each RFC community will be specifically designed to work with nature and wildlife.

RFC communities will be constructed in different configurations, depending on the region, but each will equal 1,248,000 acres in land area. If a RFC green community were a perfect rectangle it would be 30 miles by 65 miles for a total of 1,950 square miles. The cost of this acreage is calculated at an average rate of $3,000 per acre. This is based on farm and ranch acreage in the continental United States. The cost per acre will vary depending on RFC locations, but it is expected that a majority of the land for the RFC green communities will be less expensive per acre in other global locations.

Another important consideration is elevation. I do not agree with an erroneous perception that asserts: the termination of greenhouse gas emissions into earth's atmosphere will halt climate change. That popular supposition is almost as ludicrous as the continuing denial from climate change skeptics. Whether you believe in human-caused climate change or not, it is a fact that climate change is here. Whether you believe that greenhouse gas emission is the cause of global warming or not, you will never convince me that dumping billions of tons of various noxious gases into the atmosphere is not bad for humans and billions of air breathing inhabitants of this planet. See Section 20: Environmental Cleanup. [146]

Scientific evidence suggests that climate change was inevitable. However, the world's dumping of the pollutants into the atmosphere that created global warming did accelerate climate change instead of allowing nature to run its course. It is vital that the world acknowledges a future that will continue to radically change. Our planet will eventually cease to have the ice caps and ice sheets in Antarctica and in the northern arctic regions of Alaska, Canada, Greenland, Scandinavia and Siberia. Glaciers will disappear. Ocean levels could rise an estimated 216 feet when all the ice melts. Many cities and international airports will be under water and airplanes cannot lift off in air temperatures between 118°F and 126°F or above. The world will be much different after this happens. [5][6][8][131]

It is vital that everyone understands that this is a real possibility. Of course some scientists say this could take 5,000 years. But to reiterate the same phrase used by climate change skeptics: "Scientists have been wrong before." [93][112][137][139]

It is imperative, and quite prudent, that the world prepares for this outcome. All RFC green communities and Maglev transit terminals will be constructed above an altitude of 216 feet in every coastal region. And built to withstand harsher weather patterns that the rise in sea level and increased temperatures will thrust upon us, along with any tsunamis from evermore-frequent earthquake activity. Any RFC facility that's built below an altitude of 216 feet will be designed to dismantle and move to higher elevations as sea levels rise. If we are going to survive the global changes that will occur, the only real question is when, then there are many such considerations. Each RFC facility will be designed and built for the worst-case scenario.

Existing farms and ranches are invited to merge with the RFC green communities that will be constructed in their regions by executing special sale agreements favorable to the current owners. For example, if a farm in the selected region is 1,000 acres and the RFC pays the owner $3,000 per acre to purchase the property, the owner gets $3,000,000. Then becomes a RFC owner and continues to live on the property. The former owner receives $72,000 per year salary, once the RFC's operation begins, and will not have any monetary responsibility for the land. The expense of working the land will be paid for by the RFC. The former owner must continue to work on the property along with 125,415 other RFC owners. They will all share the responsibilities of the property as a cooperative in addition to the adjacent 1,247,000 acres that, altogether, forms the RFC green community.

$3,744,000,000 – 1,248,000-acre land purchase.

Start-up expenses for each RFC green community includes labor and shipping fees for the purchase of organic plants and seeds, livestock, fish, tools, equipment, A.I. robots, farm planting and harvesting vehicles, fencing, building materials, heavy machinery, etc. These are many of the expenses included in this enormous RFC budget. High-volume discounts on all purchase orders will apply. The quantities ordered, technological improvements, and construction innovation will lower these expenses for every RFC green community. All of the 1,166 RFC communities will be an integral part of the global conglomerate that includes their governing body, RFC Direct, and their organic enforcement and green technology think tank Green WALET. See Section 16: Conglomerate. [155][170]

$3,000,000,000 – Ranch and farm budgeted start-up expenses.

An assortment of varieties of fruits and vegetables will be chosen from around the world. Organic seeds will be selected from seed banks that have not been contaminated with genetically modified seeds, aka GMO. It's important an assortment of varieties of fruits and vegetables are grown and not a selected few, which are what today's farmers have thrust upon us. For instance, there are many varieties of corn and many varieties of apples that you don't see anymore. Large corporate farms have been limiting the selections for years in favor of fewer varieties of each fruit or vegetable. The benefit of the small farms is that an assortment of varieties of each fruit or vegetable can be grown instead of what corporate owned farms produce, many which specialize in mono crops of the same variety of fruit or vegetable, year after year.

RFC green communities will include thousands of small farms, each specializing in different varieties of fruits and vegetables selected by their RFC members each year. After the RFC members have selected the type of fruits and vegetables they want grown the following season and what variety of each fruit or vegetable, the RFC farmers then decide who plants which variety of fruit or vegetable in each growing season based on mandatory crop rotation. Proper crop rotation is important. One farmer may plant a variety of melon one year, then the next year plant a variety of bean. This is healthy for the land and gives the RFC members mixed varieties that they have not had the opportunity to experience from their local grocery store's produce section.

Fruit trees are different. The majority of them can't be grown in one season. Some fruit trees require up to ten years to mature before they can begin producing fruit in particular geographic regions, under various climate conditions. Many varieties of fruit trees are planted in year one. Millions of them will be selected from seeds and cuttings then planted in greenhouses and bio-domes that are designed and constructed for mass production of fruits and vegetables. Many plants can be grown in vertical conditions that will save on acreage. Some plants and fungi need to be cultivated in enclosed climate-controlled conditions, such as mushrooms. The construction of these massive greenhouses and bio-domes during the first six months of RFC surveying will allow farmers to begin seed germination and rooting of cuttings, from the selected fruit trees, in the first few months under perfect conditions. The growth rate could be sped up in the first several years after seed germination and the cuttings begin rooting. Once the trees reach a certain size they will be transferred to their selected RFC locations for climate acclimation before planting. This technique could accelerate the cultivation process by three years. So ten-year maturing fruit trees will begin producing by the end of year seven when each RFC begins production. [172]

After the fruit trees have all been transferred, other crops will be planted in the greenhouses and bio-domes, taking full advantage of the improved indoor cultivation techniques. The fruits and vegetables that will be grown in vertical and stacked positions, some hydroponically, will yield more produce using less acreage. Unfortunately, not every crop can be grown using these techniques, but the ones that can be cultivated will excel in these perfectly controlled climate conditions. [127]

Grain crops, such as wheat, rice, and oats, will require millions of acres to cultivate the quantities needed for the many products that are produced with grains that will be requested by RFC members. Gain crop acreage is added in regions best suited for grain cultivation. The millions of acres required for algae ponds, to produce biodiesel and livestock feed, will be added in regions best suited for algae cultivation. This grain and algae acreage is an addition to the global total combined acreage of the 1,166 RFC green communities. It will be purchased as a global expense and will be governed by the nearest RFC. Grains and algae are produced using the organic and regenerative guidelines followed by all RFCs. See Section 4: Water Requirements, and Section 12: Global Cost.

Livestock will be selected using the same diligence as fruits and vegetables. Livestock is used as a general term for selection, purchasing, and breeding and using it in this context collectively includes all cattle, dairy cows, goats, hogs, buffalo, poultry, sheep, etc. There are many varieties of livestock. In year one, selection of the best livestock breeds begins and will be purchased and bred to get superior progeny. This is repeated year after year until a sufficient number of each livestock breed is produced from the best linage with the best genes and the best taste. The old-fashioned way, as it was done for centuries prior to genetic engineering. By year seven the best-bred linage of each livestock species will be completed.

Fish will be more difficult than livestock to select for breeding. The freshwater and saltwater fish stocks that are purchased from fisheries will be selected for purity, but for variety, oceans will be the only resource available. Oceans today are not that clean and selecting fish for breeding purposes could pose a problem. Nobody knows exactly where fish have been. It is not the same as livestock, when you can see where the herd has been and easily determine if the land they have been on is contaminated. With fish, it may be impossible to accurately determine if the vast aquatic region they have inhabited had ever been contaminated.

Fish that migrate are obviously the hardest to accurately track to determine where they have been. Fish that don't migrate are hard as well because no one knows what ocean current carried through their aquatic region, that may have been contaminated, then continued on through while the fish stayed in their marine habitat. Shellfish are the easiest to track because they don't travel any great distance. Their surroundings must still be inspected and the water tested, but then again, nobody knows if contaminated water flowed through with the currents leaving toxic damage behind. All species will have to be tested to make sure they are pure enough to breed in the RFC fisheries, regardless of whether they migrate or not.

This process is quite time consuming for each species, involving selection, then testing, breeding, and then testing again before a breeding program can be established. Some species that migrate will be easier than others. Turtles are one of those species. Since turtles lay their eggs on land it is only a matter of catching the hatchlings for testing prior to them reaching the ocean. If the test results were negative for toxins, after testing several hatchlings from the same female, then it would be safe to conclude that the other hatchlings are also uncontaminated.

Once the land for each RFC green community has been selected, the surveying will determine the locations of all farms, ranches, fisheries, factories, ocean water desalination plants, electric micro-grids, residential neighborhoods; community, municipal and commercial buildings; fuel depot, hospital, schools, and the community infrastructure. This is the RFC configuration breakdown of this vast acreage:

```
  480,000 – Acres for crops and specialty crops
  280,320 – Acres for other crops
  249,600 – Acres for livestock for slaughter
   96,000 – Acres for livestock breeding and fisheries
  142,080 – Acres for community municipal infrastructure
1,248,000 – Total acres
```

AGRICULTURE

Crops

Minimum yields for 480,000 acres, including 46,000 flooded acres for two rice crops per year (the only grain that will be cultivated within the RFC boundaries), is figured at 12.5 people fed per acre. This is equal to 6,000,000 RFC members fed per year from agriculture production. RFC members select the type of crops grown in their RFC each year. Crops are planted to accommodate each member's ordering habits. This annual selection process is compared to the RFC ordering database that stores each member's ordering routine for current and previous years. This way the RFC farmers can estimate the quantity required of each selected crop planted the following growing season and rotate the crops accordingly to regenerate the soil and as a natural insect deterrent. (Crop rotation has been practiced for centuries and eliminates the need for pesticide. If necessary, RFCs will make their own non-toxic or rapidly biodegradable insect deterrents.) Crops, livestock, and fish are produced based on each RFC member's ordering habits.

Grocery stores currently use the same type of data techniques I am proposing by ordering the types of produce their customers want based on the daily records of their customers' buying habits. The only significant differences will be that the distributors and grocery stores will be eliminated from the current distribution sequence and each RFC member's data records of their daily ordering habits will be individually saved instead of data taken from the combined customers' sales records. This meticulous method will lower the costs for each RFC by eliminating the middlemen between farmer and consumer and helps prevent surplus crop production, which keeps RFC membership fees to a minimum.

As produce is harvested it immediately begins to lose nutrients. The longer it is stored prior to consumption the more nutrients it loses. The RFC ordering and distribution system delivers produce to members faster than current distribution methods. The cutoff time for orders is at 6:00 AM daily for the following day's delivery. Members will be assigned a picker that is responsible for compiling their orders, much like a personal shopper. These 60,000 pickers will fill the orders that come in from each of their assigned 96 members and will pick produce based on the member's criteria. For example, one member may like their tomatoes riper than another member. The pickers will be able to select produce according to each of their member's particular preferences for type, size, quality, and ripeness. Red meat, fish and poultry will also be selected according to each member's ordering specifications. Each picker will be responsible for the daily processing of certain foods, such as nut butters, vegetable and fruit juices, tomato paste and sauce, etc., in a systematic way so everything is freshly made.

There are other benefits of this type of food ordering, selection, processing, and distribution system. Food products will be thoroughly inspected, which reduces the chances of shipping a tainted product. If by chance a contaminated product is shipped, a computerized data tracking record for each order would be established that can accurately pinpoint the location of that product's source, including any added ingredients that make up that product. When large quantities of a food product are produced at one time it is much harder to track down a contaminated product. Each RFC's member's data tracking and distribution methods will eliminate wasting untainted food that is normally destroyed with a recall because the manufacturer doesn't know exactly which particular ingredient from that production batch was contaminated. Once a food product is either suspected or determined contaminated, or a member's symptoms have been diagnosed as food poisoning, then the product or an ingredient's source can be quickly identified. An immediate warning would be issued to members that ordered and received that particular food product, minutes after the diagnoses and its identification, before an epidemic occurs.

A recall would seldom happen. If a recall were to be issued, then it would not be to the scale that recalls happen today, which are because of over manufacturing of food products far in advance. When there are too many items produced in advance there isn't any way of determining exactly where and when the tainted product was made, or which fruit or vegetable ingredients from that batch were contaminated. So everything would have to be recalled. Although manufacturers put identifying code numbers and dates on products, their ability to recall a particular product successfully is limited to media that doesn't always reach every single purchaser of that product. This will not happen within the RFC system. Since each RFC member can reorder every 24 hours, then members will order smaller quantities at a time and won't feel they need to stockpile certain products in bulk for cost savings or the fear of a low seasonal availability. This will eliminate the dependency on preservatives in food products because there would no longer be a reason for longer shelf life. It also reduces each member's electricity usage by eliminating the need for a larger refrigerator or freezer for prolonged storage.

Specialty Crops

Specialty crops are crops that grow well only in particular environmental conditions that are difficult to replicate. No matter what the controlled environment is, some crops will only grow successfully in certain climates and under particular environmental conditions. Coffee is one of those specialty crops ordered in high volume. It is best to let the RFCs located in those perfect environmental regions continue to grow those particular crops in large quantities.

Since RFCs will have a Maglev transit distribution network, these crops could be distributed worldwide very quickly to fill orders placed by members of each RFC without any delay of shipments. However, there is one problem that needs to be addressed: quantity.

Unfortunately, not all specialty crops can be produced in desired quantity. For example, Kona coffee, from the Big Island of Hawai'i, can only be produced in small quantities due to the limited agricultural area. The expansion of this regional agricultural area is not possible. So these types of crops will have to be special orders with additional charges to compensate for the higher production costs and the limited availability of these products. Columbian coffee is more abundant than Kona coffee and produced in larger quantities; therefore, it will be included for no additional charges under the RFC membership agreement. Only selected products with limited availability will be an additional expense. However, the additional charges for these specialty crops will be less than what is charged for these products at today's current retail rates.

Other Crops

There will be 280,320 acres that are dedicated to other crops, consisting of guinea grass, moringa trees, and industrial hemp. Grains and algae cultivation are not included in these 280,320 acres within the RFC green community. The immense acreage that is needed for grains and algae are purchased separately in locations best suited for their crop production.

Guinea grass will be used for a processed cattle feed. Pressurized heat from an autoclave is used in a conversion process that turns guinea grass into an 85% digestible organic cattle feed. It lowers methane (CH_4) emissions and fattens cattle as fast, and with better results, than corn and grain, which are only 35% digestible. Cattle are the world's number-one producers of global methane, which is over 23 times more damaging in the atmosphere than carbon dioxide (CO_2). Feeding cattle corn and grain is the majority of the problem. Methane emits from their first stomach while burping during the digestion process. A reaction from consuming feed that is not entirely assimilated by the animals. Cattle are grass eaters and were never intended by nature to consume anything else.

Open-range grass-fed cattle does eliminate the burping issue, but doesn't fatten them as quickly, nor is the taste as desirable as cattle that are fed corn and grain. Feeding cattle processed guinea grass fattens them quickly and achieves the same desirable taste. Besides the burping and taste issues there's another problem that's associated with open-range-fed cattle: contact with other animals. Predators are not the only problem. In some African regions, cattle will frequently come into contact with water buffalo, not a predator, but a known carrier of hoof-and-mouth disease. To solve this major problem, ranchers have fenced in grazing areas for their cattle. These fenced-off areas protect the cattle, but now block the natural migration routes for the animals whose ancestors have inhabited those regions for millennia.

These issues will be addressed by each RFC. It is important that each RFC community is planned around indigenous wildlife and their migration routes. To eliminate the problem of contacting diseases from other animals, RFCs will feed their cattle processed guinea grass in large ten-acre feedlots. To overcome the overcrowding feedlot issues that are also frequently associated with diseases, the cattle will be limited to 25 per acre instead of the current inhumane feedlot methods that pack them into a small area where antibiotics must be frequently used to counteract the diseases they contact in their cramped quarters. Using this processed feed and these livestock containment methods will produce a desired beef taste by feeding cattle organically, and expose them to much less stress by treating them more humanely in these larger feedlots, with a lesser threat of contacting any sickness or disease. These ten-acre feedlot locations will change periodically as part of crop rotation to regenerate the soil. [121]

Moringa trees are used for several different purposes. They will be grown between the ten-acre feedlots as wind blocks to keep dust down and lower topsoil erosion. They are an excellent source of nutrition. The leaves are abundant with essential amino acids, vitamins, minerals, and antioxidants. It has essential anti-inflammatory, anti-diabetic, cholesterol-lowering, and cardio-protective properties. It has been used medicinally and as a food source for thousands of years. The Ayurvedic system of medicine links it to the cure or prevention of approximately 300 diseases. Their seeds work better for water purification than many conventional synthetic materials that are used today. Moringa leaves will be available to members and will be mixed with guinea grass to provide nourishment for cattle. The leaves can be harvested year-round. [9]

Industrial Hemp is a utility plant that will be used by each RFC for many different purposes including oil, rope, textiles, blankets, tent shelters, and a composite plastic material that is ten times stronger than steel. For clarification: Hemp is not the same plant as marijuana, but they are the same species. Although the medicinal purposes of marijuana are controversial, it is not compatible with hemp cultivation. RFCs are not intending to grow marijuana for medicinal or recreational purposes.

With the exception of rice, grains will not be grown within RFC community boundaries. The acreage for grain crops will be purchased separately as a global expense. Grain crops will be grown on millions of acres that may not always be suited for every RFC community location. After harvest, grains will be shipped to the nearest RFC for silo storage prior to their distribution to other RFCs. See Section 12: Global Cost, and Section 13: Global Funding. [10]

Unlike grain crops, algae could be grown in any RFC community location. Like grains, the millions of acres for algae cultivation required for biodiesel production will be purchased separately as a global expense. Algae will be grown in one-acre shallow ponds that use water discarded from the fisheries, during their routine water changes, or water pumped directly from ocean water desalination plant locations. Dried algae will be mixed with feed and will provide added nourishment for livestock as a byproduct of the biodiesel production. See Section 5: Houses, Buildings, Factories; Section 6: Electric Power and Fuel; Section 12: Global Cost; and Section 13: Global Funding. [11][12][13]

LIVESTOCK

Livestock require two separate designated locations: one for slaughter and one for breeding. There will be 249,600 acres designated for livestock slaughter. RFC animals will be free roaming, but confined within ten-acre feedlots. Each animal species will have a per-acre capacity limit to prevent any overcrowding. This lowers the possibility of transmitting a contagious disease that requires dispensing non-organic antibiotics. No animal will be fed anything but organic feed produced by each RFC green community. Veterinarians will be on duty 24 hours a day to monitor each feedlot and remove any animal that is sick or injured. This policy will certify that all animals remain healthy and are not contaminated in any way.

The necessary supply capacities for cattle, sheep, hogs, poultry and other livestock slaughtered will be based on each member's ordering demands.

There will be 96,000 acres designated for livestock breeding. This acreage will be used for select breeding purposes and will use the same caution as the livestock slaughter acreage. Only the best specimens of an animal breed will be selected for reproduction. This will assure that RFC members have the best possible lineage for consumption. Dairy and egg production will also be located on this acreage. The capacities for dairy and egg production will be based on each member's ordering demands. Veterinarians will monitor the breeding animals 24 hours a day.

The 6,600 acres for the 300 freshwater and saltwater fisheries will not be located in the slaughter and breeding areas. They will be located within the 480,000 agricultural acres and will have a distance of 20 acres between each fishery. There will be 660 acres of agriculture surrounding a single fishery. See Section 3: Aquaculture and Land-Based Fisheries.

COMMUNITY MUNICIPAL INFRASTRUCTURE

The community municipal infrastructure's 142,080 acres will be divided into residential, municipal, commercial, and industrial acreage. This will include the acreage purchased outside the RFC community's boundaries that will be used for the ocean water desalination plants and hospitals. This acreage will not include the Maglev transit distribution terminals and electric micro-grid sites. That acreage will be selected separately and is a separate global expense. See Section 9: Magnetic Levitation Transit, Section 12: Global Cost, and Section 13: Global Funding.

125,416 – Total one-acre RFC owner residential land plots
  8,984 – Acres community municipal and commercial
  <u>7,680</u> – Acres industrial
142,080 – Total infrastructure acres

There will be 125,416 acres designated for RFC owner residences, one acre per owner and their families. And an additional 8,984 acres will be for parks and recreation, school grounds for sports and other outside activities, police stations, courthouse, fire stations, building department, public venues; such as a stadium, arena, and theater; vehicle registration, roads, bike paths, non-RFC commercial stores, and various services, etc., the same as would be found in any other municipality. Every RFC green community's charter requires that their residential infrastructure include environmentally safe, sustainable, and health-improvement technologies.

The final 7,680 acres will be for RFC industrial usage. This total acreage will include 480 acres located outside the community for the 24 ocean water desalination plants that will be constructed on 20 acres each, and an additional 264 acres that will be for the four RFC hospitals and medical centers that will be constructed on 66 acres each. The other 6,936 acres, that will be reserved inside the RFC green community boundaries, will be for one hospital and medical center, a water distribution center, the RFC distribution and processing center, and the following factories and production facilities:

1) Biodegradable plastic packaging and disposable paper plant.
2) Grain mill plant.
3) Dairy processing plant.
4) Bakery plant.
5) Leather, wool, flax fiber, silk, industrial hemp fiber textile plant.
6) Livestock feed processing.
7) Oil (hemp, olive, flax, etc.) seed (vegetables, grains, fruit) processing.
8) Seed and dehydrated vegetable, grain, fruit, and moringa, processing.
9) Slaughterhouse and packaging of red meat, pork, poultry, and fish.
10) Vegetable and fruit processing, freezing, canning, bottling plant.
11) Machinery, mold, fabrication, weaving, and 3D printing plant.
12) Fuel depot: biomethane, biodiesel, ethanol, and hydrogen production.
13) Sewage treatment plants and waste management facilities.
14) Livestock methane production.

The remaining commercial and industrial acreage will be for any community approved non-RFC related businesses. RFC owners will not be prevented from starting non-RFC companies: they will be encouraged to do so. These non-RFC businesses would still be income tax exempt because they would be located within the RFC free trade zone. Non-RFC business owners' employees that reside within the community would also be income tax exempt. But, if the business were to expand beyond preset community standards and grew too large for the community, or if the RFC owner or the employees began earning more than the RFC owner salaries, then the business owner would be required to give up their RFC ownership and move their family, employees, and the business, outside of the RFC community. They would no longer receive RFC income or be exempt from income tax or sales tax. A new RFC owner-worker would be selected and would move into the vacated residence.

If the RFC owner wishes to remain an owner, then they will have the option to sell their successful non-RFC business and remain a RFC owner and stay in the community, but the business would not be allowed to stay and would be required to move out of the community along with employees that decide to continue working for that business. However, there would be a second option. If the RFC owner wished to continue to be a RFC owner and continue to keep their non-RFC business within the community, then they would be required to keep their successful business under the preset limit on size, revenue, and salaries.

As with all rules there will be exceptions and each situation will be under individual review before any action is taken. Restaurants would be a business that would be an exception because they would be supplied by the RFC and would have beneficial social and entertainment aspects.

No RFC owner or family member can separately earn more than the RFC owner salary. This policy is meant to keep RFC residents equal in pay within the community, so no one has any financial sway or takes advantage of the RFC community's free-trade zone status. If any financial manipulation is allowed, then it could disrupt the regional membership fee structure; operation production costs for food, water, and electricity; and RFC healthcare services, and could disturb the delicate balance of the nonprofit cooperative. An RFC must remain detached from capitalism's influence or any possibility of creating a rich versus poor type of division scenario within the RFC green community, which could possibly lead to corruption and voracity affecting the four requirements of human life.

Some will say this is socialism, but it is not. No one is attempting to limit the ability of anyone who wants to improve their financial status or limit anyone's innovation or ambition. Every RFC owner-worker will be encouraged to be all that they can be in life, but there are some people that find happiness in having a good-paying job, a place to live, a family, and don't have the desire to be financially better off than their neighbors. RFC owner-workers will be able to use their skillsets to do their assigned job, which would be predetermined to be the position that is best suited for them within the cooperative. And would never have to be concerned about the neighbor next door receiving a larger salary for their job. Some people will make great managers, some great farmers. Whatever an RFC owner-worker's skillsets are, they will be selected for their positions for their abilities and not for the pay grade.

Infrastructure and municipal taxes for the police, fire department, municipal government, roads, sidewalks, parks, event and entertainment venues, courthouse, etc. will be paid for by the taxes on the community's residents: RFC owners and non-owners alike. Everyone will pay a share of taxes. The same essential taxes as any other community, except most communities are supported by residential and business property taxes, which are based on the valuation of their properties. RFC community residents will pay an equal amount of taxes per person. There won't be any property tax, since all of the RFC community property will be owned collectively. Anything that is not directly related to the performance of a RFC operation will be a community and not a RFC expense.

3

AQUACULTURE AND LAND-BASED FISHERIES

There are numerous benefits of land-based fisheries. The elimination of toxins in fish farmed in the RFC fisheries will be one noteworthy benefit. The oceans have become a dumpsite for everything including mercury, sewage, garbage, trash, nuclear radiation, pesticides, toxic chemicals, etc., and naturally occurring toxins from underwater volcanic activity. This is a danger to fish health and human health from subsequent consumption. Fisheries located in the sea do not eliminate toxins and have the same problem that land-based livestock share: being raised in feedlots that are kept too cramped, thus increasing their vulnerability and lowering their resistance to sickness and disease. RFC fisheries eliminate this by farming fewer fish in one-acre pools and through selected breeding programs that strengthen their overall stamina and health. This is equivalent to breeding the best quality of livestock. [56][177]

Ocean fish farms that are raising fish species' that are genetically engineered pose another serious problem more dangerous than any land-based genetically engineered plants. If a GMO fish escapes, it could create a larger problem than contamination on land, because there isn't any way to track a genetically engineered fish in the ocean. Even tagged fish that are tracked by researchers cannot be contained, and the environmental pandemic possibilities from a single GMO contamination could be much more catastrophic in the ocean, with a much more serious consequence, than any land-based GMO.

As it is with air, viral transmission is much quicker in water than through earth. A GMO can be transmitted from one plant to another, but is restricted by the land itself, preventing it from spreading except by direct contact. Wind can aid in spreading GMO plants and pesticides by blowing them to farther-downwind locations. Birds can also carry plants to other locations. These possibilities could be prevented, but if GMOs are transmitted to an uninfected location they can be eradicated and the soil uncontaminated. Open-water, however, doesn't have any restrictions in any direction and can't be uncontaminated. The pandemic that could occur from a single exposure of a GMO product in the ocean could have disastrous consequences worldwide. [14]

AQUACULTURE

Fishermen should make a transition to farming fish rather than catching fish. However, many fishermen, boat and ship owners, and others that have made their livings working on the sea will not be able to make this transition to land-based fisheries. But there is a solution to this problem. Fishermen will be tasked with repairing the environmental damage that manmade pollutants began and help clean up the various types of debris produced by natural and human-caused disasters. This damage includes wetlands and coral reef restoration. In addition to the restoration of these affected areas, from the accumulation of decades of manmade pollution and now the rising global ocean temperature, the continued rising ocean levels must also be taken into account. It is important work budgeted for in the global RFC green community plan. Fishermen will be instrumental in helping to return the oceans, and the sea creatures that inhabit them, to a healthier condition. See Section 20: Environmental Cleanup.

LAND-BASED FISHERIES

There are 300 fisheries in a RFC green community. Each fishery is 10 acres in size and divided into one-acre pools 42 feet deep. Below each one-acre pool is a 20-foot high 880,000 cubic-foot capacity aboveground hydrogen gas storage tank used to store excess electricity. Condominiums surround the fishery on the exterior. One fishery has a total footprint of 968,000 square feet (22 acres), which includes 634,040 square feet of building and an exterior residential parking lot surrounding the fishery. This design is efficient and cost effective. See Section 6: Electric Power and Fuel.

The fishery construction technique that will be used is modeled after Chinese prefabrication techniques used to construct buildings in a matter of days instead of the months or years that conventional building construction requires. Each RFC fishery complex is composite reinforced concrete. The residences that surround a fishery complex are designed to add reinforcement to the one-acre pool walls. The exterior height of the fishery is 84 feet from the ground level to the roof of the eighth floor. [97]

The top six floors of each fishery are residential floors. Each of these floors is configured for either a one-bedroom single condominium or two floors configured for a three-bedroom condominium. Each floor will be 42 feet by 42 feet totaling 1,764 square feet. The three-bedroom configurations total 3,528 square feet. Below these residences will be two floors. The ground-level floor is for resident garage parking, for those that choose to have electric vehicles. And the second floor is a resident storage locker area with an interior walkway around the fishery from the monorail stop that is located on the non-residential end of the fishery. The elevated monorail will be constructed for a public transportation system that services the entire RFC residential and business community. See Section 7: Distribution and Transportation.

The combined footprint of the 300 fisheries will equal 3,000 acres of total water containment and doubles to 6,600 acres total footprint with condominiums and parking lot included. There is a twenty-acre distance between each fishery that will be used for agriculture. There are 660 acres that surround a single fishery, and a total of 198,000 acres that surround 300 fisheries. A combined total of 204,600 acres includes condominiums, fisheries, and agriculture.

A single ten-acre fishery has 315 three-bedroom condominiums or as many as 630 one-bedroom condominiums. One end of each fishery is not residential and does not have the top two floors above the water line. Instead there will be a two-story shed where a rolling fishing crane is stored at night. This shed is equivalent in concrete to two floors. The six floors below this shed are designed for the fishery offices, fishing control center, holding pool, monorail transit stop, and a truck loading dock on the ground floor. The rolling fishing crane is used to catch fish from each of the ten pools and then dumps them into the holding pool to be sorted before truck transport to the processing center. Only the specific fish type and the number of fish ordered by members are caught daily.

$12,524,400,000 – 300 ten-acre concrete fisheries with condominiums.
   3,300,000,000 – $12,500 floor-finish x 264,000 floors in 300 fisheries.
   3,000,000,000 – Cleaning and processing equipment for 300 fisheries.

4
WATER REQUIREMENTS

The effects of global warming and climate change have wreaked havoc on water supplies globally. Numerous glaciers have provided fresh water to people for millennia, but they are now melting at an alarming rate with the increasing temperatures from global warming. Millions of people throughout the world, particularly in high altitude mountainous regions, will cease having potable water when their supply disappears. If this were combined with a reduced snowfall, instigated by a prolonged drought, it would have a calamitous effect on the population. [8]

Severe droughts have periodically plagued regions of the world for centuries. It happens, it is an integral part of nature. But with climate change happening at an accelerated rate, many regions of the world are experiencing more droughts than what has been considered normal over the last several decades. This, combined with the increasing demands for water, is affecting groundwater aquifers and surface water; rivers, lakes, and manmade reservoirs; thereby reducing once plentiful water supplies, which these water resources have provided for centuries, to their lowest depleted levels. A municipality's ability to cope with these droughts and provide potable water to a rapidly increasing population has become a challenge. But they exacerbated the problem with serious consequences that should have been foreseen, but, until recently, was ignored. [169]

The depletion of aquifers, those endless pools of water envisioned under the ground, has been part of this problem. It can take a century to recharge a deep underground aquifer and most contemporary farms, cities and counties have been pumping them dry for decades. In sparsely populated rural regions, wells are often drilled into shallower, smaller pockets of groundwater that are not part of an aquifer. Groundwater must continually be replenished from rain or snowmelt. In prolonged drought regions a dry well is a result of consistent water usage without a recharge. As the effects of climate change continue, this consequence will occur frequently. What happens when the supply runs dry? [106][110][123]

We're at that inevitable question now and the only logical answer is ocean water desalination. Several high-capacity reverse osmosis ocean water desalination plants have been constructed during the last decade.

Perth, Australia, in 2006, was the first one to be used to supply a city with water. It is also the first one to be partially powered by a wind farm. Since then, other high-capacity ocean water desalination plants have been built in Australia, India and Israel. [96]

We cannot abandon aquifers completely, but they should only be treated as a secondary water source and not relied on as a primary water source. They should be treated as an emergency reservoir, a vital extra resource that is there for those unpredictable natural disasters. This could be a fire, earthquake, hurricane, tsunami, or a tornado destroying a water distribution pipeline. Or it could be a non-emergency situation, such as a desalination plant being shut down for servicing. These aquifers should only be used during these times of necessity. [15][125]

Water is measured in acre-feet. One acre-foot of water is 325,851 gallons and is generally rounded up to 326,000 gallons. It contains 43,560 cubic feet of water. One cubic foot of water contains 7.48 gallons.

The 480,000 acres used for agriculture crops use approximately 1,071,290,959 gallons per day, which equals 2.5 acre-feet per acre per year if supplied daily. This quantity is what is required and the high-volume amount that will be provided by desalination plants if there isn't any rain.

The other crops being grown on the 280,320-acre agricultural area are all lower-volume water-usage crops that require only half that amount, totaling 312,816,960 gallons per day, which equals 1.25 acre-feet per acre per year if water is supplied daily. Again, this is the water amount that will be required from the desalination plants if there isn't any rain. Desalination plants will be designed for these higher maximum capacities than what would be required during normal rain patterns. If droughts in agricultural regions persist due to climate change, then the desalination plants will be able to supply all water demand requirements. [127][128]

The grain crops requiring water in regions outside the RFC green community boundaries will have their water pumped in through water pipelines from the closest RFC or directly from the desalination plants as the need arises. The algae crops that require 326,000 gallons of water per one-acre pond will also have their water pumped from the closest RFC or directly from the desalination plants. For global biodiesel fuel production 1,610,946,825 acres will be required. If this acreage is divided between the 1,166 RFC green communities, then each RFC's desalination plants will need to supply 1,381,601 acre-feet of water (450,401,926,000 gallons).

The ranch acreage, 249,000 acres and 96,000 acres for livestock, requires 24 gallons of water per day for each head of cattle and various lower quantities for the other RFC livestock. For 1,500,000 head of cattle, 36,000,000 gallons per day is required. The desalination plants will supply this amount and whatever is required for the additional animals.

The total water capacity for 300 fisheries is 41,076,000,000 gallons (13,692,000 gallons per one-acre pool). An ocean water desalination plant will take approximately eight months to reach full water capacity in all 300 fisheries. Periodically, the water will need to be changed in each pool. The drained water will be used for crops, algae, or hydrogen production.

A person will generally require approximately a gallon of potable water a day for consumption and cooking. This amount, and the extra amount that is required for pets, will be provided to all RFC members. The additional water requirements for bathing, toilet flushing, washing clothes, etc.; washing vehicles, watering gardens and lawns; etc., increases the personal water requirements of each person to approximately 100 gallons per day. This additional water usage will be supplied to all RFC green community residents through the water pipeline system built into each community. Outside RFC green communities, water will be piped in to communities where RFC members reside and hooked up to existing municipal water distribution systems. In many regional locations, such as in Africa, where municipal water distribution systems are unavailable in outlying areas, RFC communities will work with the regional municipal governments to arrange potable water deliveries. Each RFC's 6,000,000 members will use an average of 600,000,000 gallons of water a day. This amount will be calculated into each RFC green community's operational output capacity requirements for their 23 desalination plants.

In regions where prolonged droughts are frequent, excess water that is produced by the RFC desalination plants will be pumped into the evaporated natural waterholes outside the community's boundaries. This replenishes the water that is essential to sustain wildlife in their habitats and helps the indigenous people within the region to survive the ever-increasing consequences of rapid climate change.

OCEAN WATER DESALINATION PLANTS

Ocean water desalination plants are modeled after reverse osmosis plants located in Israel and Australia. Capacity of the Israeli plant is 164,736,000 gallons per day. The total cost for construction was $400 million in 2013. Each RFC green community's ocean water desalination plants will have the same per-day maximum output capacity.

At 164,736,000 gallons per day, the operational output capacity of the RFC green community's 23 desalination plants will be 3,788,928,000 gallons per day. This is approximately twice what the RFC's operational demand will be for each RFC operation and their 6,000,000 members.

The additional water output over the amounts required for each RFC operation and their 6,000,000 members, will be supplied to the RFC's commercial business members, both inside and outside of the RFC green community. See Section 15: Business Memberships.

The other recipients of the additional water output within the RFC community will be schools, libraries, municipal buildings, hospital, the police and fire departments, parks and recreation facilities, event and entertainment venues, etc., and any other local community service that is not classified as a commercial business or part of the RFC operation.

One additional desalination plant dedicated to the fisheries and algae pond containment pools is required. Bringing the total number to 24 plants.

$7,893,760,000 – 24 plants: $328,906,666 each. This cost has been volume discounted for constructing the 24 plants simultaneously. B100 biodiesel turbines provide the electricity required for each plant. See Section 6: Electric Power and Fuel.

225,000,000 – 2,250 miles of water pipelines at $100,000 per mile from each desalination plant to their RFC green community's water distribution center, or to their members' locations, algae pond locations, grain crop locations, etc., outside the community. See Section 5: Houses, Buildings, Factories.

5
HOUSES, BUILDINGS, FACTORIES

HOUSES

Every RFC owner is assigned a three-bedroom house built on one acre of land inside the RFC green community boundaries. The houses will be constructed simultaneously, taking advantage of volume discounts on labor and materials and the prefabrication building techniques used to keep the cost of each house at $90,000. All electricity, gas, water, solar hot water, and sewer are provided to each house within the community. Each of the 125,416 RFC owners take possession of their homes at the end of year six, when they begin working the farms, ranches, and fisheries in preparation for the RFC to begin operation at the beginning of year eight.

$11,287,440,000 – 125,416 three-bedroom houses at $90,000 each.

BUILDINGS

Farms and Ranches

The community's barns, sheds, grain silos, greenhouses, bio-domes, grow buildings, etc., will be built in the required locations. When appropriate, these structures will integrate discarded worn-out tires and plastic bottles into their construction. Recycling tires and used plastic bottles will help keep the expenses down on construction materials for these buildings and will simultaneously eliminate these useless items from the landfills and unsanctioned dumpsites where they typically exist. Tires and plastic bottles are abundant worldwide. The plastic bottles that aren't converted to diesel fuel, road sections, or algae pond containment pools, will be recycled for building construction. See Section 6: Electric Power and Fuel, and Section 7: Distribution and Transportation. [16]

The RFC green community's water distribution center's shed will be constructed in a centralized location within the community where the pipelines from the 24 ocean water desalination plants will converge. It is nothing more than an automatic switching station where potable water is automatically routed through the community's pipelines, pumping water to the designated locations within the community. The water distribution center's pipeline shed is built on agricultural acreage.

$1,254,160,000 – Farm and ranch buildings.

Community and Municipal

Multi-purpose municipal buildings will each serve various community functions, the same as other communities. The multi-purpose buildings will include the community's offices, schools, firehouses, police station, courthouse; multi-purpose event, entertainment, and conference venues; storage facilities, etc., and will be built using prefabrication construction techniques. [97]

Schools are built to accommodate 250,000 children from first grade through high school. These schools are designed differently than traditional classrooms. The bulk of a child's studies will be online in their home. The school's buildings will be used primarily for lectures, science labs, student group interaction, and the social activities that are essential for each child's social, artistic, and physical education development in preparation for their entrance into society. See Section 18: Education.

Buildings for new non-RFC related businesses can be built on the designated RFC green community commercial and industrial acreage, but the business and the building must meet RFC preset architectural and zoning standards. If they do, then a building can be constructed by the business with a long-term lease on the land, but RFC funding will not be provided for any construction. See Section 2: Land Requirements.

$975,000,000 – Community building prefabrication construction.

Hospitals

Hospitals are modeled after Bumrungrad Hospital in Bangkok, Thailand, a JCI-accredited hospital. There are a total of five hospitals built for each RFC green community. One hospital is built within each community's boundaries and accommodates RFC owner-workers, their families, and RFC members residing inside and outside the community. Four hospitals will be constructed outside the RFC community in RFC member regions for convenient member access. Each hospital is designed with a 1,845-bed capacity and will be capable of accommodating up to a maximum of 10,000 people per hospital per day. This is equivalent to two visits per year for 6,000,000 people if every RFC member goes to the hospital twice. This also includes the combined once-a-year visit for teeth cleaning and a dentist examination and the once-a-year visit for a physical examination. These visits are mandatory as a requirement of the RFC membership and scheduled on the same day. See Section 17: Healthcare. [17]

There are 66 acres required for the preferred standard design of 28-story three-wing hospitals built inside each RFC community and in the cities and rural areas outside each community. The hospitals will be two stories each and spread out, covering 635,040 sq. ft. per floor. The 26 residential floors that are above the second floor are all two-story 3,500-sq.-ft., three-bedroom condominiums. There are 2,730 condominiums per wing. This standard hospital design will house 8,100 hospital staff. The additional 90 condominiums are for guests. Five hospitals house a total of 40,500 hospital staff per 6,000,000 RFC green community owners and members. [97]

Outside the RFC community, within the cities where limited land is available, the buildings will be purchased if the available buildings meet the hospital's requirements. This includes the housing of each hospital's employees. If suitable buildings are not available, then the required land will be purchased and the hospital will be designed and built to meet all of the specified hospital requirements. Transportable emergency medical facilities will be built for the outlying RFC members' residential areas.

Each hospital's staff will consist of 1,200 doctors and dentists, 600 specialists, 900 nurses, 600 dentist hygienists, and 4,800 employees. The staff will be required to reside on the premises during on-duty workdays as a condition of employment.

The hospitals built inside and outside the RFC community will have their own incineration process for used medical supplies and their own sewage system and water distribution system on hospital grounds in all rural regions. In the cities where hospitals and medical centers are located, and are unable to accommodate their own separate water and sewage systems, will only have the incineration process. These hospitals and medical centers will hook up to each city's existing water and sewage system already established.

$5,625,000,000 – Five hospitals at a cost of $1,125,000,000 each.

Distribution and Processing Center

The distribution and processing center is comprised of 14,031 domes, each with a walk-in refrigerator. It is essentially a gigantic market where 60,000 pickers assemble their assigned members' orders for delivery. Its centralized location in the community is where all food is packaged for delivery to residences within the community or exported by truck or Maglev flatcar to the members' regions outside of the community. The empty refrigerators are delivered to the center throughout the day where they are sanitized prior to loading the processed members' orders for overnight delivery. Once the orders are completed the refrigerators are loaded onto either trucks for local regional delivery or Maglev flatcars for shipping to a specific regional distribution terminal. From there they are unloaded onto trucks for delivery to the members' pick-up locations. See Section 7: Distribution and Transportation.

$701,550,000 – 14,031 domes at $50,000 each.

FACTORIES

The RFC factories will be constructed within the community's designated industrial areas, in multiple locations for convenient access from owner-worker residence locations. There will be thirty factories essential for RFC operation: ten factory types, three factories each type. Any factory that has an odor associated with it, such as a dairy or slaughterhouse, will be built in industrial areas that are the farthest downwind locations from the community's residences and municipal buildings. The fuel depot, sewage treatment, and waste facilities, will also be constructed in these locations.

The factories will use prefabrication construction techniques, the same as other RFC community buildings. Factories will be modeled after existing architecturally designed energy-efficient buildings, but will be engineered to use the latest state-of-the-art machinery to increase the efficiency. If there are additional specialized industries that are unique to each region, they will be considered during the RFC planning phase. [97]

The ten factory types essential for RFC operation are:

1) Biodegradable plastic packaging and disposable paper plant.
2) Grain mill plant.
3) Dairy processing plant.
4) Bakery plant.
5) Leather, wool, flax fiber, silk, industrial hemp fiber textile plant.
6) Livestock feed processing.
7) Oil (hemp, olive, flax, etc.) seed (vegetables, grains, fruit) processing.
8) Seed and dehydrated vegetable, grain, fruit, and moringa, processing.
9) Slaughterhouse and packaging of red meat, pork, poultry, and fish.
10) Vegetable and fruit processing, freezing, canning, bottling plant.

$1,998,450,000 – Thirty factories at $66,615,000 each.

INDUSTRIAL FABRICATION PLANT

This factory is for fabrication of the essential products that need to be machined, fabricated, molded, weaved, etc., for the RFC farms, ranches, buildings, and factories. Any prototype, replacement parts, or the limited production runs of any of the products required by the community, will be produced in this plant. The materials frequently used will be plastic, industrial hemp, carbon fiber, steel, aluminum, wood, etc. [116][117]

3D printing is another technology that will be used in this plant. The products and components that at one time could only be produced by machining, mold making, or stamping, are now made by 3D printers. This is a very expensive technology, but there isn't any extensive labor involved and the machines that were once necessary will no longer be required. This will significantly reduce machine electricity consumption. However, this will not totally eliminate machines for items that are not economical to print, but it could for the items that would be too time consuming to machine. This factory is for RFC community-related use only and the items produced are not sold commercially.

$100,000,000 – Industrial fabrication and 3D printing plant.

FUEL DEPOT

The centrally located fuel depot is a storage facility for four fuels. It also produces biofluids used for electric hydraulic motors that run in RFC community factories and grease for machinery. Three of the fuels stored, E100 ethanol, B100 biodiesel, and compressed biomethane (CBM), are used as power-generating fuels throughout each RFC community and distributed by pipeline. These three fuels are produced at their source and then used at the site prior to the excess quantity being pumped to the fuel depot. It doesn't make sense to waste energy trucking the required source ingredients a long distance to a biofuel plant and then wasting more energy pumping it all the way back to its origin. The RFC green communities will be designed for efficiency. The fourth fuel is hydrogen. It is produced and stored in this facility, but is not used for fuel within the community. It is only produced for airships and for energy storage. See Section 6: Electric Power and Fuel.

$100,000,000 – Fuel depot building, storage tanks, pipelines, equipment.

E100 Ethanol

Each RFC green community is comprised of many regenerative mixed farms that produce their own E100 ethanol fuel. All of it is fermented from crop byproducts and farm waste products without any dedicated land use. Many small ethanol production facilities are set up within the agricultural areas for efficiency. Excess E100 is pumped through pipelines to the fuel depot. This fuel is used in each community to power small portable generators, all-terrain vehicles, and motor tools such as chain saws and lawn mowers. A limited amount of E100 will be distributed by pipeline or transported by Maglev flatcar to RFC members outside of each community. It is either delivered directly to members or to RFC electric micro-grid (REM) converted refueling stations for distribution. E100 ethanol will be sold directly to RFC members in small quantities. See Section 6: Electric Power and Fuel.

$120,000,000 – E100 ethanol fermentation production equipment.

B100 Biodiesel

Diesel biofuel will be produced from algae oil. It will fuel the RFC green community's turbines for the factories and also fuel the truck fleet inside the community. Outside the community, it will fuel the RFC delivery and maintenance truck fleet, Maglev flatcar turbines, and the Maglev transit distribution terminal turbines; the pipeline pumps for water, E100, and B100; and the RFC electric micro-grid sites powering member residences. B100 biodiesel is not produced inside the community. It is processed at the algae ponds' location. The closest algae oil processing plant to a RFC green community will pump the fuel quantity required through a pipeline into the community's fuel depot for storage and distribution. The excess fuel is then pumped through the Maglev track pipeline to a distribution terminal's fuel depot for storage and distribution. B100 is the only fuel that is delivered for distribution in large quantities outside of the RFC green communities. See Section 6: Electric Power and Fuel, and Section 9: Magnetic Levitation Transit.

$1,000,000,000 – B100 algae oil refineries with pipeline to fuel depot.

Biomethane

This biofuel will be produced from livestock urine and manure, and the RFC green community's sewage and waste facilities. The 442 anaerobic digesters required to process the waste created by over 1,500,000 head of cattle and various other livestock, and the human waste from each RFC community, will produce the biomethane (CBM) that will be washed and compressed to run the CHP turbines, heat the autoclaves for guinea grass processing, and power the guinea grass mowers, mixers and conveyors used for the cattle feed.

    The processing of the CBM is completed at the livestock ranches and the unused excess fuel is pumped through pipelines to the fuel depot for storage and distribution. The excess electricity that is produced by the 64 1MW turbines is distributed through the community's micro-grids.

$2,652,000,000 – 442 at $6,000,000 each anaerobic digester, compressor.
    442,000,000 – 442 at $1,000,000 each autoclave, mixer, ingredient, etc.
    221,000,000 – 442 at $500,000 each automated grass mower, rake, etc.

442,000,000 – 442 at $1,000,000 each automated feedlot conveyor, etc.
4,600,000 – Organic ingredient cultivation, harvesting, mixing, etc.
76,800,000 – 64 1MW CBM-fueled turbines at $1,200,000 each.

The excess CBM fuel stored at the RFC community's fuel depot, powers an additional 1,365 200kW turbines that, along with the excess power from the 1MW turbines, supplies electricity to every home, fishery condominium, hospital, municipal building and the monorail. The fuel is distributed through pipelines to turbine locations that power individual micro-grids. And to every house and fishery residence for an optional gas flame cooktop, instead of an electric cooktop, which will be the only non-electric appliance. See Section 6: Electric Power and Fuel, and Section 7: Distribution and Transportation.

$327,600,000 – 1,365 200kW CBM-fueled turbines. 273MW x $1.20/watt.

Liquid biomethane (LBM) is a clean burning liquid fuel that has the same high quality as liquid natural gas (LNG). Though the BTUs are lower than CBM and CNG, the trucking is easier. CBM and LBM both have their advantages and disadvantages. Converting CBM to LBM is better for shipping by truck or flatcar and then converting it back to CBM is better for burning because the BTUs are higher. This conversion will only be made if transport is required. See Section 6: Electric Power and Fuel. [18]

Hydrogen

There are three common production processes to make hydrogen.

1) The water electrolysis method is the simplest procedure. It consists of two interconnected columns of water that each has an electrode wired to a DC power source. Hydrogen is released into a gas at the negative electrode and oxygen is released in the water column with the positive electrode. As long as the columns containing the water are constantly refilled the process continues making pure hydrogen and oxygen, which are the only two elements produced when electricity is applied. The process requires almost 33kWh of electricity to produce a gallon of hydrogen.

2) Injected steam into methane gas. This is a more complex method than the first method of water electrolysis. It is lower in cost, but produces an almost equal amount of carbon dioxide. The process takes methane gas and injects steam to produce hydrogen and carbon monoxide. The carbon monoxide gas is then combined with oxygen to produce the waste byproduct carbon dioxide.
3) The liquid methanol electrolysis process. Developed by NASA, this procedure is complex, but uses less electricity than water electrolysis. Using methanol that is passed through a chamber divided by a thin special membrane with electrodes of opposite polarity located on each side, the hydrogen ions pass through the membrane and move toward the negative electrode, leaving behind the waste byproduct carbon dioxide. The level of carbon dioxide is much lower than method two.

Water electrolysis is the cleanest method even though it requires the most power. But using methane and methanol, procedures two and three, produces carbon dioxide, which is one of the greenhouse gases the world is attempting to reduce. Though they both use less electricity for each process, the additional energy and equipment required to produce methane and methanol doesn't make procedures two and three any less expensive than process one. There is a new process that uses methane to produce hydrogen without any carbon dioxide emission, but, again, the methane must still be produced. Making a clean burning fuel to make a clean burning fuel negates the advantage of this method. Using renewable energy generation from solar or wind produces electricity for free, except for the initial cost and required maintenance of a solar system or wind-driven turbines, consequently making water electrolysis the best option.

Hydrogen will be produced for airships and electricity storage. Airships use hydrogen for lift and are refilled at the fuel depot. Power is stored in the fuel depot's hydrogen tanks and in the concrete tanks built under the fisheries. It is then released during the peak demand periods. See Section 6: Electric Power and Fuel, Section 3: Aquaculture and Land-Based Fisheries, and Section 10: Rigid Airships. [19][20][21][22]

$120,000,000 – Hydrogen production equipment.

SEWAGE TREATMENT AND WASTE MANAGEMENT FACILITIES

Sewage Treatment Plants

Sewage treatment plants will convert sewage to fertilizer pellets for non-edible agriculture. These plants also produce methane that is compressed into CBM fuel. The CBM will then be used to fuel each treatment plant's turbine power generators. Any excess CBM produced is pumped through pipelines to the RFC fuel depot for RFC community distribution.

Waste Management Facilities

There will be multiple waste management facilities constructed in every RFC community in close proximity to each residential, commercial or industrial location. They will have a smaller footprint, using fewer acres, compared to typical landfills and will recycle waste by using it for energy production instead of letting it decompose in landfills over time.

Unrecyclable waste that is delivered to a facility will be disposed of using two procedures. The first procedure uses anaerobic digesters to convert waste to methane, then compresses it into CBM to fuel turbine generators. Excess gas is pumped to the fuel depot. The second procedure uses an incinerator. The heat created from the incineration process will be used to make either hot water for distribution to nearby community locations or combined with solar thermal to make steam for their turbine power generators. These procedures will reduce waste odor by efficiently disposing of all unrecyclable waste immediately after delivery.

Incinerator water-heating furnaces will be used in conjunction with individual solar hot water systems to provide unlimited hot water. Electric water heaters will be supplied for backup only. Using the water-heating furnaces in nearby waste facilities avoids long-distance trucking of unrecyclable waste and will reduce the pipeline distance for hot water distribution.

Any exhaust gases that are produced by the incineration process are captured and compressed, then pumped through pipelines to the fuel depot where the CO_2 is filtered out from the other gas elements. The CO_2 is then pumped to the desalination plants where it will be combined with hydrogen to produce a synthetic substitute natural gas (SNG), which is comparable to the CBM and CNG fuels. The remaining gas elements will be disposed of using filtration and storage techniques that eliminate any greenhouse gas emission into earth's atmosphere. See Section 6: Electric Power and Fuel.

RFC recycling requirements for glass, aluminum, paper products, plastic, metal, wood, electronics, appliances, etc., will limit the amount of unrecyclable waste disposed of at waste facilities. All recyclable waste will be shipped to various approved recycling facilities. Green waste from tree and hedge trimming, garden maintenance, etc., will be used for ethanol or composting. Plastic will be shipped to RFC conversion facilities to be converted into diesel fuel, or manufactured into either road sections or algae pond containment pools. See Section 6: Electric Power and Fuel, and Section 7: Distribution and Transportation. [62][67][180]

$320,000,000 – Sewage treatment plants and waste management facilities.

6
ELECTRIC POWER AND FUEL

The world has a power problem and we all know it. The issue is not that we need or use too much power. The issue is that we are not mass-producing it in a clean, renewable, and sustainable way. The methods of electric power generation that we have continued to rely on have been inadequate or dangerous to the environment, wildlife, and humans.

The big challenge to solving the global energy problem is providing enough clean renewable and sustainable electricity to not only meet global consumption today; 25,551,300,000MWh per year – total 2017 world power generation; but bring the world up to United States standards, the largest per capita power consuming country, which is approximately four times more. If we're going to eliminate famine and poverty, then equality with industrialized countries will have to be met. The global consumption rate would then require 102,205,200,000MWh per year. However, this does not include any natural gas and heating oil consumption, which are used in appliances, water heaters, furnaces, etc. If gas and oil BTUs were converted to electric and the future replacement of fossil-fueled vehicles with electric vehicles was considered, then the global electricity consumption rate would be considerably more for BTU consumption equality with the United States. [11][23]

This is a massive global goal and beyond the scope of the RFC green community plan. The RFC goal is to produce enough power to supply all electricity within each community and to each ocean water desalination plant, hospital, Maglev distribution terminal, fuel depot, etc., that are located outside each community's boundaries; and to RFC members to meet their personal, home, and vehicle requirements. It also includes powering the Maglev transit network.

This massive quantity of electricity cannot be generated by one method alone. It must be produced from several different sources. In order to make the best decision on what types of power generation best suits the necessities of world demand and environmental protection, all current and future conceived electric power generation methods must be considered and examined.

POWER-PRODUCING TECHNOLOGY

Electricity is produced by one of three methods: turbine generators, fuel cells, or solar photovoltaic. Turbines are designed for a specific type of energy that turns the generators to produce electricity. No mechanical movement, no electricity. Fuel cells require a specific type of fuel for an electro-chemical reaction to produce electricity without any mechanical movement. Solar cell photovoltaic converts sunlight directly to electricity.

Steam, Water, Ocean, Gas, Fuel, and Wind Powered Turbines

Steam turbines are used in nuclear, geothermal, solar thermal, ocean thermal, and coal and petroleum steam-driven power plants. Water-driven turbines are used in hydroelectric power plants and are submerged in flow-controlled rapid water current. Underwater water-driven turbines are powered by either river or ocean current, and ocean tides and waves. Expansion turbines are powered by a rapid expansion of released high-pressure gas. Fueled combustion turbines are used in power plants burning petroleum products; oil, diesel, natural gas; and the renewable fuels biodiesel, biomethane, and ethanol. Wind-driven turbines are powered by the wind.

Turbines produce most of the electricity powering the planet. These massive generators of electricity are also capable of producing heat. Combined Heat and Power (CHP) plants are in many locations throughout the world. In addition to the massive quantities of electricity that are produced by these CHP turbines an additional amount of heat energy is produced, which is distributed separately. [98][99]

Generators

Generators produce electricity the same as a turbine generator, except a generator is powered by a fueled motor instead of a turbine engine and typically has a lower electricity output. They can vary in size and can be either stationary or portable. They are most frequently used for back-up electricity during power outages at homes and businesses. These portable generators are also used at locations where electrical power is either not available or inadequate for power demand. Generators are available with a variety of power outputs and motor types to choose. Motor fuel options to consider include gasoline, propane, diesel, and natural gas. The BTUs of a fuel determine the quantity consumed per hour.

Fuel Cells

Fuel cell technology has been around for decades, but the primary focus has been DC power for vehicles not for mass power generation. One company has perfected a fuel cell modular system that generates AC power for large-scale purposes using either natural gas or methane biogas for fuel. When using natural gas it emits limited CO_2, but when using methane biogas, its emissions are zero. The modules are scalable so any increased demand in power can be added easily in 50kW increments. A 250kW net output AC power module weighs 38,800 pounds. And the efficiency of the system is rated between 52 and 60 percent. In addition, fuel cells produce heat during the electro-chemical reaction. Capturing the heat generated does increase the fuel cell's efficiency rating in a CHP system. [24][25][26]

Solar Photovoltaic

Solar cell photovoltaic technology converts sunlight directly to electricity. See **SOLAR**. [27][28]

THE GRID

There are electric power grids throughout many countries – the United States, Canada, China, countries in the European Union, etc. – but for the purpose of the following explanation I am using the United States.

The electric grid in the United States is nothing more than a vast network of over 300,000 miles of transmission lines that are connected to substations transmitting electricity from more than 9,000 power plants. Seven percent of the power generated is lost in the transmission alone (line loss) and many of these power plants on the grid are designed for one purpose only: peak power demand. While the largest power plants continue to produce power around the clock, these satellite power plants are designed to supply power only during the peak hours of the day, mostly during the evening hours when the largest amount of residential power consumption occurs. These satellite power plants operate less than six hours a day and are the dirtiest, least efficient, and most expensive, environmentally disastrous power generation plants on the planet.

What does that have to do with the grid? Nothing really. It is only an explanation of how our power distribution has progressed since 1882, when Thomas Edison first flipped on the switch, turning on the power generators in New York City, which began our ascension into the era of centralized electricity generation distributed by what has become known as "The Grid."

It was a momentous engineering feat in the 20th century, but has outlived its usefulness in the 21st century with an emerging renewable energy market that is much better suited for micro-grids and stand-alone systems of power generation.

I do not understand why there is such a perceived necessity to have a national electric grid. If a disaster strikes, then power is lost to everyone. When micro-grids and single-user systems shut down it only affects those people on that particular micro-grid or single-user system, not everyone. There would be no rolling blackouts or the need to transfer power from one region to another. The Grid is basically obsolete in today's world of individual power generation technology. With micro-grids, no one entity is in control deciding who gets power and who doesn't. If a micro-grid requires more power, then an additional power-generating device is installed. If the power demand is great enough, another micro-grid is constructed. There won't be any reason to sell power through The Grid to a community that requires it and then charge a premium to that community's consumers for the convenience. This common procedure is old and outdated. [132][133]

Was The Grid necessary? Yes. Have we outgrown it? Definitely. However, some people don't see it that way. The Grid has always been a one-way street. Power plants generate the power and consumers receive the power. But today the renewable energy industry has emerged and now the one-way street power grid has been mandated to become a two-way street. Electric power companies have been compelled to purchase power from business and residential customers' wind and solar systems that are, unfortunately, inconsistent power generators. [100]

This has created havoc and given people a false impression that if they sell renewable electricity to the power company they are doing their fair share to lower greenhouse gases emitted into the atmosphere. That is not how it works. The power company's turbines are not dispatchable, which means they cannot be slowed down when electricity surges happen and then sped back up when power surges cease. Without an electricity storage system, the additional electricity goes unused. So to exploit this misperception even further, there is a movement to upgrade The Grid to a Smart Grid. Which is a two-way street that allows buying and selling of electricity generated by the power company and business and residential customers that purchased their own electric power generating systems, essentially competing with the power company.

Digital smart meters that provide real-time electricity data will replace the old meters with the spinning disks. Smart Grid technology aspects will add computer intelligence and digital communications to the electricity distribution network, thus permitting smart houses and smart businesses to be in constant communication. Business and residential electricity consumers will be able to track their energy usage over their smart phones or computers. Plug-in electric vehicles, appliances, water heaters, etc., will communicate over the Smart Grid, striving to find each system's most efficient way to operate. This strategy will revolutionize how electricity is generated, transported, and consumed throughout the United States. An estimated total of $1.5 trillion will be spent upgrading The Grid to a Smart Grid infrastructure over the next quarter-century. Whether a new Smart Grid is constructed that maximizes efficiency and reduces waste or The Grid as we know it is upgraded to a Smart Grid doesn't matter. It is still an obsolete single electrical system technology that uses a sophisticated, but vulnerable, computer network that can be hacked and disabled.

All RFC individual electric micro-grids will incorporate Smart Grid technology features into their infrastructure, but they will not have the vulnerability characteristics for disabling an entire electrical network from a single hack.

ELECTRICITY STORAGE

Energy Storage Solutions (ESS), the fancy words for a 21st-century battery, has been the primary problem for many renewable electricity-generating technologies. [100][140]

Battery

Whether batteries are lead-acid, chemical, sodium-nickel, nickel-ion, lithium-ion, flow, etc., they all have their recycling and recharging issues. Lead-acid batteries are not recyclable while sodium-nickel batteries are recyclable. Flow batteries have corrosion problems, but the electrolyte is reusable. Regardless of the recyclability they all have a limited lifespan determined by their rated number of charge and discharge cycles. Solar photovoltaic (PV) panels have a lifespan of thirty years, but even the best batteries will need to be replaced every ten years or less. Flow batteries could have a 20-year lifespan, but until the internal corrosion issues are resolved it could be expensive to maintain. Reliable electricity storage in batteries continues to be the problem with renewable energy technology. [29][151]

Another problem with ESS is the components that are used to manufacture each unit. Continually using finite natural resources is not a solution, only a bandage applied that masks the problem. A total solution to any electricity storage must be one that is long term and doesn't require an overabundance of mined minerals for a battery with a short lifespan. Even though sodium (Na) is the 6th most abundant mineral on the planet and vanadium (V) is 20th, nickel (Ni) is 24th, and lithium (Li) is 33rd, they are still finite natural resources that must be mined from the earth's crust. It is more acceptable if a battery has a long lifespan and is recyclable, but a total waste of resources for any short-term product without having full recycling capability.

In a micro-grid or single-user system, combinations of ESS and solar PV or wind energy can supply renewable electricity to communities and facilities without heavy investment in transmission and distribution. Other renewable power generation, such as biodiesel and biomethane turbines, can be coupled with ESS to cover peak demands and provide emergency power during power outages. ESS systems have their uses on communication transmission networks to resolve temporary overload issues on those networks caused by consistent irregularity in electric power supply. But the big picture of solving the same issues on a global scale is not economically feasible, or more importantly, as previously stated; too short of lifespans, recyclability, and a tremendous toll on natural resources.

Flywheel

A flywheel battery is a mechanical apparatus for electricity storage. A rotating mass moving at extremely high revolutions per minute stores kinetic energy, then instantly converts it to electric energy when it is needed. One company's patented flywheel battery has 175,000 full-depth charge and discharge cycles. This is stunning compared to the 1,000 to 10,000 cycles for other ESS. It has a much lower per-cycle cost and will last decades longer than any current battery on the market, while being almost maintenance free. It always spins in the same direction and has the ability to instantaneously shift back and forth between charging and discharging modes. This flywheel battery system can be configured for a range of applications from 100kW to multi-MW systems and is scalable from one flywheel up to several hundred. There aren't any environmental problems with either processing or disposal. Its long lifespan negates the amount of natural resources required for manufacturing compared to other ESS.

Flywheel batteries could be a possible solution for large-scale commercial wind farms and solar PV flat-panel power plants, but are not a solution for either the small-scale solar PV or wind power installations used for business or residential electricity generation. They have been used very successfully when combined with other ESS. And they have a long lifespan. However, their electricity discharge time is short and they have one potential problem: They can explode if the rotating mass moves out of alignment. This will only happen if the bearings fail or a major earthquake disrupts its rotation. The best solution for either possibility is to install them underground. If an explosion does occur, then they will be contained and prevent any injury to people or damage to structures in the nearby vicinity. [30][31][32]

Pumped Hydro Storage (PHS)

This PHS method does not store electricity, only water. The process uses excess energy during the non-peak demand hours to pump water up to a storage reservoir constructed at a higher elevation. When peak power demand is required the reservoir releases the water and the pressurized flow from the elevated water turns the hydroelectric turbines, generating enough electricity to meet peak demand. This method works best with a renewable energy source, such as solar PV or wind that produces excess electricity during the non-peak demand hours to pump the water to the reservoir location.

If there is an elevation high enough to build a reservoir without any environmental impact, enough excess water available, and close to the peak power demand location, then this can be a reliable solution for renewable energy use without an additional ESS. [33][126]

Compressed Air Energy Storage (CAES)

This method uses excess energy, produced during the off-peak hours, to compress ambient air and store it under pressure in a large underground cavern. Similar to PHS, this process doesn't store electricity. It only stores the fuel, in this case heated pressurized air. When peak power is required, air is released and rapidly expands in an expansion turbine to generate electricity. [34]

Liquid Air Energy Storage (LAES)

This method uses excess energy, produced during the off-peak hours, to cool air down to -196°C, turning it into a lower-volume liquid stored in a low-pressure insulated tank. When it is released, the liquid air returns to its normal gaseous state by rapidly expanding in an expansion turbine to generate electricity. [148]

Hydrogen Energy Storage

Hydrogen has a stable chemistry. A small amount can store a significant quantity of electricity for extended periods. And it has a long discharge time. It is a practical solution for storing intermittent energy generated by solar systems and wind farms. Using the water electrolysis process, hydrogen gas is produced and then compressed to more than 2,400 psi to reduce its very light volume. It is then stored in aboveground-pressurized tanks, or in an underground cavern, where it is charged with electricity. The electricity is discharged during peak energy demand. See Section 5: Houses, Buildings, Factories. [22]

Substitute Natural Gas (SNG) Energy Storage

SNG has a greater energy storage capacity than hydrogen and a much longer discharge period. It is gasified using coal, which requires mining. Though there are ample coal reserves to last 250 years, it doesn't store enough excess energy to justify the cost of the gasification process and the cost of coal mining. It is an expensive alternative to hydrogen energy storage. But producing a synthetic SNG by combining hydrogen, that has been produced using water electrolysis powered by free excess energy, and CO_2 captured by artificial trees and from diesel-fueled turbines and engines, makes synthetic SNG production economically viable as a CNG substitute that can store excess energy prior to distribution. [22][35][114] [119][120]

NUCLEAR

Nuclear fission reactors produce steam for their plants' turbines and can generate enough electricity to meet global demands, but at the same time the dangers are extreme with lasting environmental consequences if a disaster strikes. There are many warranted concerns that include nuclear accidents, nuclear proliferation, terrorism, drought, radioactive waste disposal, etc. These issues have all been debated for decades. Then there are unpredictable natural disasters, such as the March 2011 tsunami that struck the Fukushima nuclear plant in Japan, triggered by a massive earthquake. While still leaking radioactive isotopes over four years later, low-level radiation reached the Pacific Ocean shores of North America, justifiably raising people's concerns about continuous exposure to low-level radiation. [36]

The world needs to contain global warming, cease polluting, and resolve global energy instability. We should invest only in the optimum sustainable energy options, but nuclear fission energy, though it could solve the global warming problem, is too great of an environmental risk. It may not be an appropriate solution to global warming. But I do agree that any realistic plan to reduce reliance on fossil fuels, thereby reducing the emissions of greenhouse gases, may need an increased use of nuclear energy. Numerous environmentalists have examined how nuclear power could be used to resolve our global warming issues. However, according to the Union of Concerned Scientists, over one-third of the U.S. nuclear power plants have suffered safety-related incidents during the previous several years. The nuclear regulators and power plant operators need to significantly improve their safety inspection techniques to prevent these devastating possibilities. These aging nuclear plants should be retired and replaced with safer nuclear technology.

Once-through cooling systems, that are common in older nuclear plants, have come under investigation for the possibility of damage to the environment. Wildlife could become trapped inside the cooling systems and killed and the increased water temperature of returning water could impact local ecosystems. Enforcement of regulations has required some older power plants to replace existing once-through cooling systems with new recirculation systems.

Severe droughts, which have become increasingly more common due to climate change, are a danger to nuclear power plants. Most plants are constructed on the shores of oceans, lakes, and rivers. They rely on submerged intake pipes to draw billions of gallons of water for use in cooling and condensing steam after it has turned each plant's turbines. If water for cooling becomes scarce then reactor output must be reduced to a lower operating power or shut down for safety. [101][123][125]

Over 250,000 tons of extremely high-level radioactive nuclear waste is temporarily stored worldwide. All decommissioned reactor sites currently store their spent fuel in concrete and steel silos, which require maintenance, monitoring, and guarding. Nuclear waste storage prevents reuse of these sites. There continues to be an international consensus on the advisability of permanently storing high-level nuclear waste in deep underground repositories, but the Waste Isolation Pilot Plant (WIPP) in the United States is the only site currently open. Another site is scheduled to open in Finland in 2020. [37][77][147]

Though nuclear fission energy is risky it does present a reasonable argument that nuclear power is in fact the only feasible way of meeting the growing global demand for electric power; while also eliminating the greenhouse gas emissions responsible for global warming. They are cost competitive per watt to build and over the last decade have improved their efficiency to nearly 90%, which makes them the most efficient power plants in the world. Regardless of the cost, efficiency, and improved safety features, they are still vulnerable to terrorism and natural disasters. It is a dangerous solution to global warming. [38][39][87][88][89]

A Small Modular Nuclear Reactor (SMR) is an emerging nuclear fission technology that will be available in the 2020s. SMRs have a 50MW output and can be manufactured in a factory, then assembled on site in half the time of a traditional nuclear plant. This will remove a significant cost issue associated with plant construction. They will be scalable, to add additional SMRs if required, and they won't need to be shut down for refueling. This will increase their efficiency to 100%. And they cannot be hacked into or melt down. [150]

Though SMR technology is an improvement in nuclear fission, nuclear fusion is by far a better solution for nuclear power. However, the fusion technology advancements, though very promising, are not close enough to perfection to be able to rely on the technology as a current energy solution. The International Thermonuclear Experimental Reactor is now being built in southern France. In ten years it could prove nuclear fusion works. And if they succeed, the fusion process will be affordable and won't produce any high-level radioactive waste. There won't be any problems with finding fuel and there won't be any byproduct produced that could be converted into nuclear weaponry. Smaller, less expensive SMR power plants and future fusion power plants can be constructed to eliminate The Grid in favor of micro-grids. [7][40][41]

HYDROELECTRIC

Hydroelectric power has proven its reliability for electricity generation, but has caused extreme environmental damage. Moving water is needed to turn the massive turbines that generate the electricity, but damning up rivers for power generation and water supply has severely impacted the biological, chemical, and physical properties; and their fragile streamside environments. These issues, combined with the location and necessity to transmit the massive power generated long distances, make this type of power generation impractical and obsolete in this new world of renewable energy and micro-grid solutions.

Dams block fish migrations. And they trap the sediments critical for maintaining physical processes and wildlife habitats. They prevent the natural maintenance of a river's deltas and barrier islands and the coastal wetlands. The transformation from a free-flowing river ecosystem to an artificial manmade reservoir habitat changes the water temperature and chemical composition. It dissolves oxygen levels and physical properties, which are not suitable for the aquatic plants and animals that evolved over time within a given river system. Reservoirs are known to host invasive species, which can include snails, algae, and predatory fish. This further undermines a river's natural ecosystem. Recent studies of major rivers have demonstrated that the river sediment and nutrient flow drive biological processes far into oceans, which prevents ocean acidification. But rivers also serve as a carbon sink for CO_2, which causes acidification.

Manmade reservoirs, especially those in the tropics, are significant contributors to global greenhouse gas emissions, on par with the aviation industry, which is about 4% of global emissions. The larger dams have led to extinctions of fish and other aquatic species. They are responsible for the disappearance of birds in floodplains; an enormous loss of forests, wetlands, and farmland; and the erosion of coastal deltas. These are only some of the impacts. There are many more. [42][125][181]

WIND

Wind-driven power is a renewable source of electricity generation, but is also inconsistent. Wind propeller-driven turbines have a very high center of gravity requiring exorbitant foundation costs, maintenance is frequent and expensive, and no matter what their size is, they are dangerous for birds. Vertical wind turbines aren't a danger to birds, but haven't been accepted as an alternative to propeller-driven turbines. Several alternative energy companies have previously experimented with different types of unusual-looking wind-driven generators, but without great success. [29]

There is one wind technology that is exceptional: a bladeless wind turbine. It is a conical tower without any gears or bearings and works on vibration. This reduces the expense of manufacturing and maintenance considerably. It has a small footprint and is noiseless. It is respectful of nature and doesn't require any energy or training to use. The bladeless wind-driven generator produces electricity with very few moving parts. It is designed to reduce the auditory and the visual impact of traditional propeller-driven turbines, and with a low center of gravity there is no need for massive foundations for installation.

To generate electricity, it relies on the oscillation of its carbon fiber mast to move a series of magnets located inside the mast. It isn't as efficient as a propeller-driven turbine, but this is offset by fewer moving parts. With the turbine effectively floating on top of the magnets, it significantly amplifies the oscillation and eliminates friction, and the need for lubrication. Compared with traditional wind turbines it has greater than a 40% reduction in its carbon footprint. This system offers direct economic advantages using the swirling motion of the wind and not the direct force needed by propellers. It generates energy from a repeating pattern of vortices produced as the air separates to pass by the body of the structure itself.

The wind's movement improves their efficiency by generating more power per square foot. And the towers can be grouped much closer together than propeller-driven turbines. The 9-foot-high tower produces 100W and the 40-foot tower has a 4kW output. The best wind propeller-driven turbine will collect approximately 50% of the winds energy, but these bladeless wind turbines are close to 40%, as was demonstrated in the company's wind tunnel. Though 10% may seem substantial, it is insignificant when compared to the 51% lower costs of manufacturing and the 53% lower operational costs under the costs of the propeller-driven turbines. The operational lifespan estimates are between 32 and 96 years. [43][44][45]

SOLAR

Solar photovoltaic (PV) flat-panels, concentrating photovoltaic (CPV), and solar thermal are three renewable technologies of solar electricity generation, but are inconsistent. When the sun doesn't shine, electricity isn't generated. Solar hot water is included in this sub-section. [29]

Solar PV Flat Panel

Solar PV panels have a very low efficiency rating of only 15 to 22 percent. Additionally, there are several key environmental issues associated with solar PV panels that completely negate the fact their use can help in the elimination of fossil-fueled power plants that emit greenhouse gases.

Solar PV panel manufacturing, shipping, and installation have their negative effects on the environment. From the mining of high-purity quartz to the toxic chemicals used in the manufacturing process, to the mass-energy required to produce the panels, the manufacturing of solar PV panels is not green. Over 50% of the world's solar PV panels are manufactured in countries that have the worst environmental records. Shipping the panels to other countries requires significant quantities of fossil fuel for a low-efficiency product. Panel installation uses billions of gallons of water at large-scale 200MW to 500MW power plants for dust control during their construction, and millions of gallons annually for panel washing. And this is in addition to the water usage for the cooling, chemical processing, and air pollution control during manufacturing. [46][47]

Then there is the problem of recycling worn-out panels, which is an extremely important issue that needs to be addressed. As the volume of solar panel installations grow worldwide, recycling processes are being implemented in anticipation of the industry's continued global growth. The solar industry recognizes the importance of an entirely sustainable recycling program. PV panels are currently required to pass the Toxicity Characteristics Leach Procedure (TCLP) test. The PV panels that pass the test are classified as non-hazardous, but have yet to be regulated. [48]

These are important issues and are a perfect example of green gone wrong. Why is a product that drains our planet's natural resources; is energy intensive and toxic to manufacture, has a short lifespan, a low efficiency, and doesn't have any environmentally sufficient small-scale or long-term batteries, or a proven recycling program; being advertised and marketed as a sustainable and energy-saving solution? I understand a person's desire to help contain global warming, as they simultaneously reduce their own electricity consumption cost without compromising their lifestyle. Which is an aspiring goal to have, and one that has proven ideal to exploit.

The sun's energy output is free and some energy companies have innovative technologies for collecting solar energy without using solar PV panels. But regardless of the type of technology used, the efficiency is still low. Until efficiency is raised, and environmental and storage issues resolved, solar panels are not a viable global energy solution. The current PV panel technology does lower the electricity usage that is supplied from The Grid at each individual residential or business installation, but that is only during the days the sun is shining. Though the sun rises every day, which is more reliable than the wind, it may be obscured by heavy cloud cover, which would significantly reduce any solar system's power output. Or the solar panels could be constantly covered with a thick layer of dust from particulates in the air, requiring frequent washing to increase their power output.

Solar Hot Water

Solar hot water, though not an electricity producer, is an energy saver. There are several systems on the market that are energy efficient, low in manufacturing costs and environmentally safe and non-toxic to produce. Water heating is very energy intensive. Solar hot water systems should be required in all new residential and commercial construction.

Concentrating Photovoltaic (CPV)

These solar photovoltaic systems were developed in the 1970s and use either lenses or curved mirrors to focus sunlight onto a small area of photovoltaic cells, increasing the amount of electricity generated. They are more efficient than flat-panel solar PV. CPV systems normally magnify the sun's radiation about 500 times, but a High-Concentration Photovoltaic Thermal (HCPVT) system magnifies radiation 2,000 times to achieve an 80%-power efficiency rating. It is a CPV parabolic dish that increases the sun's radiation while also producing fresh water. This CPV system uses a dense array of water-cooled solar chips that convert 80% of the sun's radiation into energy. It is not suitable for a rooftop system. It is 33 feet high, weighs 10 tons, and is 47 square yards. It generates 12kW of electrical power and 20kW of heat. The mirrors concentrate the sun's energy onto the chips to produce electricity. Reaching temperatures of 1,500° Celsius the chips are cooled to 105° Celsius with a water radiator system to stay within their operating temperatures. The dense array of photovoltaic chips, mirrors, and the electrical receiver are encased in a large transparent plastic enclosure to protect the system.

The radiator system, filled with the water that does the cooling, also creates hot water that can be used in either a space heater or for air conditioning through an absorption chiller. The hot water can then be used in the desalination system and creates potable water by passing water through a membrane filter. This system has the ability to produce as much as 350 gallons of water a day. The parabolic dish has 36 elliptic mirrors made of a very thin 0.2-millimeter thick recyclable aluminum foil with a silver coating. The photovoltaic chips are mounted on micro-structured layers that continuously pump piped water within a few tenths of a millimeter from each chip. This absorbs the heat and draws it away much more effectively than passive air-cooling. The estimated operating lifespan, based on the system's current design, is up to 60 years with periodic maintenance. Depending on the system's environment, the required replacement of the foil and the elliptic mirrors will need to be done approximately every 10 to 15 years and the photovoltaic cells every 25 years. [49][50]

Solar Thermal

The only type of solar system that makes sense on a commercial scale is the thermosolar plant in Seville, Spain. It has the ability to operate during the day while storing heat energy when the sun is at its peak. It then delivers that stored energy when the market demand is at its peak, during the evening hours when the electricity consumption rate is highest. This type of system changes everything in the solar power market. It can operate around the clock and the fuel and electricity storage cost is zero.

This Concentrating Solar Power (CSP) plant uses a central tower receiver combined with Molten Salt Energy Storage (MSES), which basically operates like a battery for heat not electricity. The innovative advancements in this technology are its molten salt receiver, its heliostat aiming system, and its control system. Using millions of square meters of mirrors, not solar PV panels, to reflect sunlight on a single point on the plant's central tower, it heats the fluid flowing past it up to several hundred degrees Celsius and uses that superheated fluid to drive an industrial-scale steam-driven turbine. Each heliostat has a reflective mirror surface of 1,184 square feet and follows the sun using two motors that have built-in programmable logic controllers that recalculates and readjusts the heliostat's position 15 times per minute. The motors move the heliostats every 4 seconds. The plant consists of 75 acres of solar heliostat aperture area with a power island and 2,650 heliostats, each with a 1,300-sq.-ft. aperture area, distributed in concentric rings around the 460-foot-high central tower receiver. The total land use for the heliostats is 480 acres.

The system uses two tanks of molten salt; one 290°C cold salts tank and one 565°C hot salts tank. Storing energy by heating molten salts to a temperature of 565°C allows the plant to reliably and flexibly produce steam to generate electricity. It routinely produces electricity up to 15 hours without sunlight. This storage capacity makes it manageable so that electricity can be supplied based on power demand. The plant has been able to supply a full day of uninterrupted power using thermal transfer technology. And has achieved continuous power production by operating 24 hours per day for a record 36 consecutive days, which is a result that no other solar plant has ever attained. Total operation is 6,450 hours at full capacity per year, a 75% capacity factor. For comparison: Hoover Dam's hydroelectric power plant in Nevada, USA, has a capacity factor of almost 23% and China's Three Gorges hydroelectric plant has a capacity factor of about 50%.

Unlike earlier parabolic trough designs that uses thermal oil as a receiver, and then transfers the heat to the molten salt solution, this plant achieves higher efficiencies through direct heating of the molten salts. A significant benefit of this type of heat storage system is the reduced load variations and increased turbine efficiency. Because of the large storage capacity, the output capacity of the turbine is smaller than what would be expected from a plant of its size. The thermal receiver has a 120MW solar thermal capacity, but the power generation plant only uses a wet-cooled two-cylinder reheat steam turbine, with a 19.9MW capacity, that supplies an annual production of 110GWh. Which is enough electricity to supply power to 27,500 homes. This power plant has been operational since May 2011 and has exceeded all expectations. At $12.41 per watt, the total cost: $247,000,000. [51]

The Crescent Dunes solar project in Nevada, another Molten Salt Energy Storage (MSES) system, is a less expensive and a larger 110MW plant, but it only has a 10-hour heat energy storage capacity. At $6.70 per watt, the total cost: $737,000,000. [52]

There are three negative aspects of CSP besides the cost per watt:

1) The concentrated sunlight from the mirrors is a danger to birds if they fly into the intense sunbeam between the mirrors and central tower.
2) The high-capacity factor is variable depending on the number of sunny days per year, requiring a back-up power source.
3) The amount of water required to clean the mirrors, tower receiver, and to cool the steam turbines.

OCEAN WAVE, TIDAL, AND CURRENT FLOW

There are thousands of terawatt-hours of incident wave and tidal energy produced along coastlines of continents each year. Tapping less than a quarter of this potential energy could readily produce as much electricity as the entire world's hydroelectric power plants.

Waves travel across the oceans and are produced mostly by winds blowing across the surface. Their arrival time at a wave power plant is more predictable and consistent than wind arriving at a wind farm. Tidal energy is different. It is driven by the gravitational pull of the moon and is very predictable hundreds of years in advance. Ocean currents are also predictable.

Using ocean wave, tidal, and current flow technologies for power generation have been quite successful, but they still have major problems to overcome. Environmental drawbacks include changes to shoreline and marine ecosystems that affect fish populations, the increasing frequency and intensity of hurricanes, and the ocean environmental damages that could occur. Other drawbacks including underwater maintenance of the power turbines, and only coastal regions will significantly benefit from these technologies, are issues that need to be considered. Geography also influences the power generation potential of each of these technologies.

Waves

Global wave energy is the best between 30° and 60° latitudes in either hemisphere, but the potential appears to be the greatest on the western coasts. There are three types of power generation. One type uses floats, buoys, or a pitching device to generate electricity using the rise and fall of the ocean swells to drive their hydraulic pumps. The second type uses an oscillating water column (OWC) device that generates electricity at the shoreline using the rise and fall of water within a cylindrical shaft. The rising water drives the air out of the top of the shaft, powering an air-driven turbine. The third type is a sort of tapered channel or overtopping device that can be located onshore or offshore. This device concentrates waves and drives them into an elevated reservoir where power is then generated using hydroelectric turbines as the water is released.

Tidal

Several global sites could be developed and potentially produce thousands of megawatts of electrical power from tidal technology. The common model for tidal power plants involves erecting a tidal dam across a narrow bay. As the tide flows in or out it creates uneven water levels on either side of the dam. Water then flows through hydroelectric turbines to generate power. The difference between high and low tides must be at least 16 feet for this type of model to be feasible. La Rance Station in France, operating since 1966, is the world's largest tidal dam and has a rated capacity of 260MW.

In 1984, the Annapolis Tidal Station in Nova Scotia, Canada, blocked an inlet of the Bay of Fundy to produce 20MW. In the United States in 2012, TidGen did it in Cobscook Bay, near the town of Eastport, Maine, without an underwater dam, which is a better solution for fish. Taking advantage of the twice-a-day 20-foot tides that rise and fall, they used a submerged lawnmower-shaped machine that produces 180kW.

These technologies, which have been developed over the last several decades, can capture the power of waves and tides and convert them into clean electric power. However, both wave and tidal energy are, unfortunately, variable in nature.

Several other models for tidal power plants have appeared in recent years including tidal lagoons, tidal fences, and underwater tidal turbines, but none are currently operating on a commercial level.

Current Flow

In the United States, a company in Florida has developed a system they call an Ocean Energy Turbine, which takes advantage of the ocean current of the Atlantic Ocean's Gulf Stream by tapping its ocean currents to produce energy by using a slow rotation with more torque. There are several other tidal power companies that have also developed tidal current turbines. These turbines will be placed in offshore underwater farm locations between 60 and 120 feet deep, where the strongest of the tidal currents flow in excess of 5 mph, spinning the turbines and generating electricity. These tidal current turbines are designed smaller than wind-driven propeller turbines and generate more electricity in a lesser area footprint. Verdant Power Company installed a pilot-scale turbine in New York City's East River in December 2006, and hopes to eventually install a 10MW tidal farm at the site.

These wave and tidal energy power plants can generate electricity without emitting greenhouse gases, although the environmental impact to the marine ecosystems and ocean fisheries is uncertain. But several projects are currently undergoing environmental studies and monitoring. The New York East River tidal turbine pilot project uses a $1.5 million sonar system to monitor the environmental impact on the waterway's fish populations. Minimal impacts on the marine ecosystems, fishing and other coastal activities should be quite low with careful placement, and gentle sitting, of each unit on the ocean floor or riverbed.

However, with the ever-changing effects of global warming, the ocean levels will begin to rise and may affect wave, tidal and current flow predictability. The continuing global changes we are experiencing on our planet could permanently change the ocean current flow directions in the future. Rising ocean levels will affect wave turbine systems and tidal turbine spinning effectiveness and will require repositioning to shallower waters. Wave, tidal, and current flow technologies are currently more expensive per megawatt than other technologies. [53][173]

OCEAN THERMAL

Ocean thermal energy technology has recently emerged from its nascent stage and is not as variable as waves, tides, and ocean current flow, but is much more expensive.

In 2015, the first closed-cycle ocean thermal energy conversion plant in the United States was brought online and is currently harvesting clean energy. The Ocean Thermal Energy Conversion (OTEC) produces electricity by pumping large quantities of both cold, deep seawater and warmer surface seawater to run a power cycle. Though the construction cost is high, OTEC has an advantage over other turbine power plants.

One of the major problems with solar and wind power is that they are inconsistent. Power plants aren't designed to slow down and speed up their turbines to accommodate surges in solar and wind power being added to the grid. OTEC turbines are dispatchable, which means that their steam turbines can be sped up or down quickly to accommodate fluctuating peak power demands or intermittent power surges produced from solar and wind. It eliminates the need for any type of ESS.

Located on the Big Island of Hawai'i, this 105kW plant cost $5 million to build. It is the largest OTEC power plant in the world. This demonstration plant shows promise, but the technology is limited to tropical coastal regions where deepwater is available. It may be a possible combined power and ocean water desalination plant solution in some tropical locations. However, a 1MW OTEC power plant is scheduled for construction in Japan, which is not a tropical region. But if successful, it could expand operational requirements to non-tropical coastal regions that have cold deep seawater off their coast and warmer surface seawater.

A proposed $1.3 billion 100MW OTEC power plant, that will be constructed on a stationary platform that is secured on the ocean floor, is planned for a yet to be selected offshore location. This design will save a considerable amount of money by assembling the 10-meter diameter fiberglass pipe on the platform as it is lowered down into deepwater, instead of the expense of running the pipe from deepwater to shore. An electrical cable will be run from the power plant to shore, which is more economical. [54][55]

GEOTHERMAL

Geothermal has proved to be a consistent source of electrical energy in 24 countries, producing over 13,800 megawatts of electricity in 2017, but the inherent dangers and the drilling cost can outweigh the benefits. Power plant construction could adversely affect land stability. And well drilling alone can cost up to $5 million per megawatt with output costs between $0.04 and $0.10 a kilowatt-hour. It is only viable in certain geographical regions. If there is a problem, it could trigger earthquakes and the release of deadly toxins into the atmosphere, or spill toxic water onto the land that can eventually leach into groundwater. It uses a minimal amount of land for the power plant's footprint and the freshwater usage is quite low compared to coal, oil, and nuclear power plants. It doesn't require any fuel. However, an explosion could be environmentally devastating to the surrounding region. This risk, as well as the high cost, is not an acceptable power alternative. It also produces approximately 45 kg of CO_2 equivalent emissions per each megawatt-hour of generated electricity using steam turbines. The countries that are currently leading the way in geothermal power are Iceland, the United States, Mexico, Italy, Indonesia, and the Philippines. [56]

A new project is being launched that requires deeper drilling than the traditional geothermal depths. The deeper drilling into the earth's crust will harness super-heated water, currently dubbed Dragon Water, in massive quantities. To release Dragon Water, deep-drilling techniques developed by the oil and gas industry must be used. The drilling depth will be between two and three kilometers. The real technical challenge is the intense heat and pressure released. It will turn steel brittle and wreak havoc on drilling electrical equipment. Engineering tools will have to be developed that can withstand these conditions. No one has yet managed to control the forces of this extreme pressure and high temperature.

This type of supercritical water isn't a liquid and isn't steam. It is in a physical form incorporating both, taking on entirely new properties. From these depths there is ten-times more energy than from a normal geothermal well. A massive amount of energy could be tapped anywhere in the world – if successful. It is also possible that within this supercritical water, valuable minerals can be transported to the surface. This could be an additional source of revenue to offset some of the costs of drilling and the expense of the specialized equipment. [57][58]

The depths of the drilling will vary from country to country due to variations. And as with traditional geothermal drilling, the possibility of earthquakes, explosions, and environmental damages, along with CO_2 emissions that are produced per megawatt-hour, will still be a concern. The enormous costs associated with deep drilling and manufacturing the specialized equipment required for this quest will continue to be an issue, making this an interesting idea, but far from an ideal sustainable energy solution.

MAGNETIC

The myths of magnetic perpetual motion generator motors sounded too good to be true and has proven to be just that, too good to be true. Although a company in Hungary was able to get a prototype to actually work, breaking a law of physics or, as they explained it, enhancing the law not breaking it. Unfortunately, once in third party testing, they were unable to keep the test units running for more than a week as various parts continued to fail because of the resonance frequency produced by the high revolutions per minute while in motion. A problem that could not be resolved no matter how they strengthened the failed internal parts.

FUELS

Fossil Fuels

Coal, petroleum, and natural gas have been the world's dominant fuels for a century. Coal and petroleum are non-renewable natural resources used in power plants worldwide. Their continued usage as fuel to meet the global daily energy demand for electricity and for motor vehicles, small engine tools, and machinery are responsible for the massive release of billions of tons of greenhouse gases into our planet's atmosphere, vastly contributing to our global warming and the rapid acceleration of climate change. One coal-fired power plant emits 1,001 kg of CO_2 per megawatt-hour when not coupled with carbon capture and storage (CCS). Until the closure of fossil-fueled power plants, and the elimination of fossil-fueled vehicles, small engines, and machinery, the emissions will continue; and so will the massive environmental damage produced by mining, drilling, hydraulic fracturing (fracking), refining, pipelines, and the global ocean shipping of these fossil fuels. [119][120]

Natural gas (CNG) and heating oil are both used for more than electricity generation and motor vehicles. Natural gas and heating oil pipelines are run in many regions to supply the fuel needed for business and residential heating and appliances. Furnaces for space heating use heating oil or natural gas. CNG is used for cooktop stoves and ovens, water heaters, clothes dryers, etc. With the exception of gas cooktops for open-flame cooking, all natural gas and heating oil furnaces, natural gas appliances, and fossil-fueled vehicles, can all be completely replaced with electric furnaces, appliances, and vehicles. Natural gas for the cooktops with open-flame burners could be replaced with SNG or CBM fuels and could be distributed through pipelines, fuel delivery by truck to on-site storage tanks, or portable tanks filled at a drive-up fuel depot.

Substitute Natural Gas (SNG)

There are several reasons to produce SNG. It can be used in existing gas pipelines that are installed in many regions instead of using CNG. It can also be used as fuel for vehicles, generators, and fuel cells. SNG is energy intensive to produce from coal and though there are enough coal reserves throughout the world to last another 250 years, it is much too expensive to produce when there are other gas alternatives available. There isn't any reason to continue mining coal. As with petroleum, coal's usefulness as a fuel is coming to an end. [35]

Producing hydrogen by electrolysis using free energy and adding CO_2 to make a synthetic SNG is less energy intensive, and less expensive, than coal conversion. SNG nearly quadruples the BTUs of hydrogen as a fuel and also stores energy. The millions of hours of electricity generation to produce enough hydrogen to make the synthetic SNG is not a waste if the energy is free and coal mining isn't required.

If a hydrogen production storage facility is constructed within an ocean water desalination plant, built at a consistent wind location, and then bladeless wind turbines are used to power the facility's hydrogen production, then hydrogen could be produced around the clock. Using artificial trees to capture CO_2 from the atmosphere and capturing CO_2

from fossil-fueled turbines and engines, then combining the CO_2 with hydrogen, a synthetic SNG could be produced. Using this procedure is a comprehensive solution to our global warming by converting CO_2 into a useful natural gas fuel replacement. Then for synthetic SNG distribution, a pipeline will be run from the hydrogen production facility's fuel depot to an existing natural gas pipeline installation. [22][114][119][120]

Hydrogen

Hydrogen is our planet's lightest, most basic, and abundant element. It is usually combined with other elements. So returning it to its pure state for use as fuel requires a considerable amount of energy. This is unfortunate, but a necessary process if pure hydrogen is to be separated from the other elements such as H^2O. Unfortunately, it takes more electricity to separate hydrogen from oxygen than its rated BTUs after the separation process is completed. [22]

There are much better fuels that require less energy to produce and have much higher BTUs for the same quantity of fuel. The only real advantage over fossil fuels is its zero greenhouse gas emission from an automobile's internal combustion engine. It only emits water from the engine's exhaust. However, economically, hydrogen is much better suited for electricity storage or supplying lift in airships than for use as a clean-burning fuel. But if it is injected into a small lightweight fuel cell, which has no moving parts, it will produce DC electric power that's ideal for powering an electric vehicle. See Section 5: Houses, Buildings, Factories; and Section 10: Rigid Airships. [82]

Alcohol

E-100 ethanol is a grain-derived alcohol and methanol is a cellulose-derived alcohol. These are the only practical alcohol fuel types. Ethanol is more economically feasible to produce on a smaller scale and methanol is more economical to produce on a larger scale. Both of them are clean-burning fuels. Gasoline-fueled portable generators, small motor tools, all-terrain vehicles, etc., are easily converted to ethanol. Methanol will not be produced in RFC green communities.

Each RFC will produce all its own ethanol from crop byproducts and crop waste without dedicated land use. A United States Department of Agriculture (USDA) report in 2004 found that ethanol nets as much as 67% more energy than it takes to produce. [59]

Biodiesel

B100 biodiesel is a renewable fuel, but is not clean burning. Made from a diverse mix of feedstock including recycled cooking oil, soybean oil, and animal fats, it is the first, and only, Environmental Protection Agency (EPA) designated Advanced Biofuel to reach over one billion gallons in commercial-scale annual production in the United States. B100 can be used in existing diesel engines without any modification. It will be stored for distribution in all RFC green community fuel depots and will be used for electricity generation and transportation. But the currently dominant soybean crop feedstock and other feedstock will be replaced with algae. [60][61][175]

Algae Oil Production

Algae oil produces high-quality biodiesel. Grown in manmade shallow one-acre ponds, it is feasible to supply each RFC's energy needs with this one crop. There are several other agricultural crops that can be used to produce B100 biodiesel, but algae are the best option, out-producing soybean oil, the nearest crop competitor, by thousands of gallons and millions of BTUs per acre. Algae also soak up massive amounts of CO_2 from the atmosphere and they leave behind biomass suitable for cattle feed and biochar. Unlike other crops, algae can be grown in climates not appropriate for agriculture, such as non-arable desert. The algae pond containment pools will be manufactured using disposable plastic waste. The same process used to manufacture roads.

Another process being used cultivates algae in enclosed tanks. It produces oil by injecting sugars into the tanks to promote algae growth without sunlight. It has worked successfully. However, algae do not soak up atmospheric CO_2 using this method and the cost of the tanks for the quantity of oil required will be more than the open-air pond cultivation technique. And with high cost and without CO_2 removal, this process is not as desirable. However, if this process proves to produce more algae oil at a lower cost than the open-air pond technique, then algae pools could be enclosed as tanks. See Section 7: Distribution and Transportation, and Section 20: Environmental Cleanup. [62][63][64][65][66]

There are other methods for atmospheric CO_2 removal besides algae. RFCs intend to use 200 million artificial trees, which are specially developed reusable filters that are coated with a resin. Installed in algae pond locations at a ratio of one tree per eight ponds; each tree captures CO_2 a thousand times more efficiently than living trees. When combined they capture a potential 10 ppm of CO_2 per year. The resin filters must be soaked in water to release the CO_2. After release, the CO_2 is compressed and pumped to a water desalination plant's hydrogen production facility, where it is combined with hydrogen to make synthetic SNG. [22][114]

Global crop production on 1,610,946,825 acres growing 100 tons of algae per acre and extracting 50% oil would produce 9,750 gallons of B100 biodiesel per acre for a total of 15,706,731,543,750 gallons per year. Out of this total acreage, 633,420,671 acres would produce enough B100 biodiesel to power the flatcars for the global Maglev transit network and fuel the trucks and farm vehicles for 1,166 RFC green communities. The remaining 977,526,154 acres would produce enough B100 biodiesel to fuel up to 64 million 200kW turbines generating a net output of 187kW each at 86°F, producing a combined worldwide total of 104,839,680GWh of electricity per year to power the global REM networks. Which is four times the 25,551,300GWh of electricity the world produced in 2017. [23]

The technology is currently available to produce B100 biodiesel on this massive scale. To eliminate any of the already ultra-low emissions from selected diesel turbines, they can be altered to capture the exhaust, thereby preventing any greenhouse gases from entering the atmosphere.

The expenses and labor for B100 biodiesel, CO_2, and synthetic SNG production will be paid for from the Maglev transit commercial distribution revenue, from synthetic SNG sales revenue, and from the electricity portion of each RFC member's yearly membership fee. See Section 9: Maglev Transit, Section 13: Global Funding, and Section 15: Business Memberships.

There is only one negative aspect of algae cultivation. If the algae are released into a regional water ecosystem it could have a cataclysmic effect. Algae have previously developed in rivers, lakes, reservoirs, oceans, and coastal areas. In some regions of the world it's destroyed ecosystems.

Plastic-To-Diesel Conversion

Fishing boats require massive quantities of diesel fuel to operate. Many of the fishermen, and boat and ship owners will be contracted for ocean cleanup and aquaculture restoration. Some of the plastic recovered by this global fleet will be converted to diesel fuel in a process known as pyrolysis. This conversion process can produce as much as 200 gallons of diesel fuel per 2,000 pounds of discarded plastic waste, providing mass quantities of fuel for these vessels as part of the global contract.

Plastic processing plants will be constructed in coastal locations near harbors and marinas. There will be more plastic recovered from the oceans than will be required for diesel fuel. The remaining quantities will be used for manufacturing algae containment pools or road sections for road construction in RFC green communities and in rural regions where roads do not currently exist. After the millions of required algae pools are constructed and the RFC and regional roads are built, road sections will continue to be manufactured for new roads and for section replacement when existing road sections are either damaged or worn out.

Selling diesel fuel commercially and road sections for non-RFC road construction creates a nonprofit revenue stream that pays for the plastic-to-diesel and plastic-to-road section conversion processes. This processed diesel fuel is not used by RFC trucks or Maglev flatcars. The annual United States on-road diesel market is approximately 40 billion gallons. That equals about 200 million tons of plastic waste. The diesel fuel market for fishing vessels is comparable.

The pyrolysis procedure currently costs approximately $40 to $52 per barrel. This is comparable with the algae to diesel production costs, but unlike algae biodiesel these plastics are made from petroleum and are returned to their original petroleum state. So emissions are still a concern. This issue can be resolved the same way it will be with B100 biodiesel fuel: by capturing the exhaust of trucks and fishing vessels and using the CO_2 for synthetic SNG.

If disposable petroleum-based plastic production was stopped and biodegradable plastic products were manufactured instead, the 96 trillion tons of discarded plastic waste that has been discarded every year for the last several decades will still exist. This conversion process can help eliminate petroleum-based plastic waste until it is eradicated and the planet is relieved of this major environmental disposable waste problem. See Section 5: Houses, Buildings, Factories; and Section 7: Distribution and Transportation. [62][67]

Biomethane

Biomethane is an abundant renewable resource that is produced from the decomposition of human waste, sewage, agriculture, and vegetation. It is also produced from anaerobic digesters processing the urine and manure from various types of RFC livestock. It is classified as a greenhouse gas that is over 23 times more damaging than CO_2 when it's released into the atmosphere. But when it is captured, it can be used as a clean-burning fuel. Both compressed (CBM) and liquid (LBM) biomethane have the same high-quality standard as both compressed (CNG) and liquid (LNG) natural gas fuels. See Section 5: Houses, Buildings, Factories.

Biomethane, methane, and natural gas are similar. Biomethane is a mixture of methane and CO_2. Methane is CH_4. When referenced in this book they are considered interchangeable. Natural gas is methane with hydrocarbons such as ethane and propane, and also contains gases such as nitrogen, helium, sulfur compounds, CO_2, and water vapor.

FUEL COMPARISONS

A British Thermal Unit (BTU) is defined as: The amount of heat energy required to increase the temperature of one pound of water by one degree Fahrenheit. Knowing the BTUs of a fuel will establish the fuel's consumption rate for mechanical power conversion to electricity. All fuel types are different. For example, propane aka liquefied petroleum gas (LPG) produces 91,420 BTUs per gallon. Gasoline produces 120,388 BTUs per gallon. It will take more LPG per hour to run a generator than it will for gasoline. The BTUs of each fuel determines the quantity of fuel consumed in the same amount of time.

When selecting a fuel to use it is important to consider the BTUs of that fuel, but equally as important to consider the exhaust emissions of that fuel. For example, gasoline and CBM have similar BTUs, but only gasoline emits greenhouse gas into the atmosphere and CBM emits none.

And there are other considerations. How much energy does it take to produce each type of fuel? Gasoline is produced from petroleum, which is extracted from oil wells drilled into the earth's crust. Then it is refined into gasoline. Biomethane can be produced from livestock urine and manure that has been processed in an anaerobic digester and then compressed to make CBM. Whatever fuel production method is used, it has to be calculated for both the monetary cost and the environmental cost of producing the final fuel product.

Electricity is also measured in BTUs and the amount of electricity produced per fuel type consumed in each method of power generation, or in non-fueled power generation such as solar, wind, or hydro, the cost per kilowatt-hour (kWh) to operate and maintain must be considered.

The following fuels are measured in either gallons or pounds. Fuels measured in pounds are converted to gallons for BTU comparison with other fuels listed. Lower and higher heating values are the BTUs produced at each end of the fuel quality spectrum. Electricity is measured in BTUs per kWh for comparison. [109]

Gasoline E10
1.0 US gallon Gasoline E10 is 112,114 – 116,090 BTU (lower)
1.0 US gallon Gasoline E10 is 120,388 – 124,340 BTU (higher)

Diesel (low sulfur)
1.0 US gallon Diesel is 128,488 BTU (lower)
1.0 US gallon Diesel is 138,490 BTU (higher)

Biodiesel B100
1.0 US gallon B100 is 119,550 BTU (lower)
1.0 US gallon B100 is 127,960 BTU (higher)

Propane (LPG)
1.0 US gallon LPG is 84,250 BTU (lower)
1.0 US gallon LPG is 91,420 BTU (higher)

Natural Gas Compressed (CNG)
1.0 US pound CNG is 20,160 BTU (lower)
1.0 US pound CNG is 22,453 BTU (higher)
(GGE) Gasoline Gallon Equivalent = 5.66 pounds
1.0 US gallon = 5.66 GGE x 20,160 BTU = 114,105.6 BTU (lower)
1.0 US gallon = 5.66 GGE x 22,453 BTU = 127,083.98 BTU (higher)

Natural Gas Liquid (LNG)
1.0 US pound LNG is 21,240 BTU (lower)
1.0 US pound LNG is 23,726 BTU (higher)
1.0 US gallon = 3.55 pounds x 21,240 BTU = 75,402 BTU (lower)
1.0 US gallon = 3.55 pounds x 23,726 BTU = 84,227.3 BTU (higher)

Biomethane Compressed (CBM)
1.0 US pound CBM is 20,160 BTU (lower)
1.0 US pound CBM is 22,453 BTU (higher)
(GGE) Gasoline Gallon Equivalent = 5.66 pounds
1.0 US gallon = 5.66 GGE x 20,160 BTU = 114,105.6 BTU (lower)
1.0 US gallon = 5.66 GGE x 22,453 BTU = 127,083.98 BTU (higher)

Biomethane Liquid (LBM)
1.0 US pound LBM is 21,240 BTU (lower)
1.0 US pound LBM is 23,726 BTU (higher)
1.0 US gallon = 3.55 pounds x 21,240 BTU = 75,402 BTU (lower)
1.0 US gallon = 3.55 pounds x 23,726 BTU = 84,227.3 BTU (higher)

Ethanol E100
1.0 US gallon of E100 is 76,330 BTU (lower)
1.0 US gallon of E100 is 84,530 BTU (higher)

Methanol
1.0 US gallon Methanol is 57,250 BTU (lower)
1.0 US gallon Methanol is 65,200 BTU (higher)

Hydrogen
1.0 US pound Hydrogen is 51,585 BTU (lower)
1.0 US pound Hydrogen is 61,013 BTU (higher)
1.0 US gallon = .592 pounds x 51,585 BTU = 30,538.32 BTU (lower)
1.0 US gallon = .592 pounds x 61,013 BTU = 36,119.696 BTU (higher)

Electricity
3,414 BTU per kWh

The nine fuels used for comparisons for electricity generation – gasoline, diesel, biodiesel, propane, natural gas, biomethane, ethanol, methanol, and hydrogen – have been taken from the Alternative Fuels Data Center Fuel Properties Comparison Chart. [68]

RFC POWER SOLUTION

Every component of each RFC green community relies on electricity. All the methods of power generation and their fuels have been considered. The result of the examination and the conclusion attained is based on cost, efficiency, environment, and wildlife protection. Although it was not within its intended objective, the recommended solution for power generation is not limited to RFC green communities and their members. It is much more extensive. It is a global solution and, along with this RFC plan, can be accomplished in seven years.

The RFC green communities will be powered by a combination of fuels and technology. RFC selected CHP turbines use a patented air bearing that increases their longevity and with the proper maintenance should have a minimum of forty years of service before any overhaul is required. [69]

CBM-fueled CHP turbines power homes, municipal buildings, conveyor systems and monorail. At ranch locations, CBM is produced, compressed and used to power 1MW CBM CHP turbines. Excess CBM is pumped through a pipeline to the RFC fuel depot for storage and RFC distribution. CBM will be supplied to each RFC community residence for gas flame cooktops.

B100-fueled CHP turbines will power the commercial buildings, factories, fisheries, and all pipeline pumps. The B100 will be produced at algae pond locations and pumped through a pipeline to the fuel depot for storage and RFC distribution.

E100 ethanol fuel will power all small motor vehicles, tools, and equipment. E100 is produced at agriculture locations and used in those locations. Excess E100 is pumped through a pipeline to the fuel depot for storage and RFC distribution. It is then shipped or pumped through a pipeline to a Maglev transit distribution terminal's fuel depot for RFC member distribution.

Bladeless wind-driven turbines will be installed throughout the RFC communities. The electricity generated will be used to produce the hydrogen for airship refilling and for electricity storage. The hydrogen is stored in pressurized tanks at the fuel depot. Electricity that is stored in pressurized tanks is used during windless periods to continue hydrogen production.

2MW hydrogen electrolyzers, the size of a shipping container, are installed next to bladeless wind turbines in various rural locations within the community where REM power is unavailable. They will provide the electricity to the outlying community ranch and agricultural areas. [22]

Hydrogen will be produced at each fishery and stored in their ten aboveground 880,000 cu. ft. concrete pressurized tanks under ten one-acre fishery pools. A tank has a 680MW storage capacity totaling 6.8GW for each ten-acre fishery. A RFC green community's 300 fisheries store 2,040GW of electricity that will be released during peak power demands.

Fuel cell modules, fueled by CBM or SNG, will be used for each RFC communication datacenter. Back-up hydrogen-stored electricity will be used when the system is down for maintenance or the fuel supply is disrupted.

B100-fueled CHP turbines power the REM sites, Maglev flatcars and distribution terminals, RFC hospitals and medical centers, and the pipeline pumps outside the community boundaries. The B100 produced at algae pond locations will be pumped through a pipeline to the Maglev transit distribution terminal's fuel depot for storage and distribution.

B100-fueled CHP turbines and bladeless wind turbines power the algae oil ponds and B100 biodiesel processing plants; and the ocean water desalination plants, hydrogen production facilities, and the pumps for fuel storage and distribution. Hydrogen will be compressed and stored in aboveground-pressurized tanks built underneath the desalination plants and on the facility's grounds.

Hydrogen is only used to refill airships, store electricity, and produce synthetic SNG. CO_2 and synthetic SNG are also stored in these tanks. CO_2 is captured from diesel-fueled turbines and engines and by artificial trees installed at algae pond locations and attached to airships. It is pumped to the desalination plants through pipelines or it is delivered by airships when refilling occurs. The CO_2 is combined with hydrogen to produce synthetic SNG. From there SNG is pumped through a pipeline to natural gas pipelines for distribution and also used to store electricity at the desalination plant. The SNG stored electricity is used for hydrogen production during windless periods.

Local refueling stations will be converted to REM sites and used to supply electricity to RFC members' residential neighborhoods and buildings, and to business and commercial parking lots that have electric outlet installations for electric vehicle recharging. They will continue to supply the fuels; B100 biodiesel, synthetic SNG, and E100 ethanol; on a limited bases for vehicle refueling. All B100 biodiesel-fueled engines will require exhaust-gas emission capturing.

Small engines – lawn mowers, chain saws, blower motors, ATVs, motorcycles, etc. – will be required to convert to E100 ethanol. The CO_2 emissions from these small fossil-fueled engines contribute significantly to global warming. So their conversion to E100 ethanol is mandatory for total worldwide elimination of manmade greenhouse gas emissions.

Fossil-fueled vehicles will be phased out as local refueling stations are converted to REM sites, eventually eliminating fossil fuel availability. RFC members get electricity included in their memberships so conversion to electric vehicles will eliminate their vehicle fuel expenses. Fuel is not included in memberships so any purchase of B100 biodiesel, synthetic SNG, or E100 ethanol will require payment. [160]

Automobile manufacturers should equip their electric vehicles with both primary battery power and secondary SNG fuel cell power for backup if the battery runs low. The fuel cell will act like a reserve tank when recharging is required. This should give vehicles enough extra miles to get to a recharging location. This is necessary in colder climates where batteries don't have the electric charge longevity that is typical in warmer climates. Hopefully future battery technology innovation will rectify this problem. Although hydrogen was never intended for fuel, it will be supplied for hydrogen-fueled vehicles in the colder climates where batteries have the shortest charge duration. [70]

Synthetic SNG fuel will be supplied by each water desalination plant's production facility. The quantities of synthetic SNG fuel will be limited compared to electricity and must be purchased when refueling. Converted refueling station REM sites will supply synthetic SNG fuel for vehicles, but electricity will be the primary energy supplied. Synthetic SNG fuel is only for backup. Electricity is included in RFC memberships, thus electric vehicle recharging is more economical than refueling.

The conversion process to an all-electric society will take many years. Property owners will not be forced to replace their reliable CNG appliances. Synthetic SNG will continue to be supplied as long as the demand is there. However, gas is not included in a RFC membership. So if an owner wants to stop paying for gas in addition to the membership fee, which includes electricity, then it is up to the owner to convert their property to electric. Eventually gas will be phased out and the conversion will be mandatory. But that could be many years in the future and not before nuclear fusion power, or an equivalent non-polluting sustainable power source, is prevalent and installed at REM sites.

As it is within RFC communities, many RFC members outside the communities prefer cooking with a flame. So gas cooktop conversion will not be required. It will remain the only gas appliance after conversion to all electric is completed. This reduces the quantity of gas consumed, lowering the demand for synthetic SNG.

Until synthetic SNG can be produced in mass quantities, CNG and heating oil must continue to be supplied by oil and gas companies to meet user fuel demands for their gas appliances, water heaters, furnaces, etc., and heating oil furnaces until they are replaced with electric. Heating oil will be phased out before natural gas. Conversion to CNG furnaces will then be required. CNG will be phased out once synthetic SNG can meet the demand.

We should not abandon existing non-fossil-fueled technologies that are being used successfully in particular regions, especially areas that are isolated and difficult to supply.

Iceland's geothermal technology should continue to flourish. The thermosolar plant in Seville, Spain, is successful and B100 turbines or hydrogen electrolyzers for backup would complement the system. Solar thermal systems should continue to be built in appropriate rural regions if the danger to birds can be eliminated. Ocean wave, tidal and current flow underwater turbines will continue to generate electricity until the effects of rising ocean levels terminates their electricity output. OTEC is expensive and locations are limited.

Environmentally friendly and successful PHS systems should remain operational. Hydroelectric production from river dams should be terminated once B100 turbine REM sites are operational and fulfilling electricity demand requirements. Once hydroelectric is terminated and desalination plants are in operation, river dams should be torn down and natural ecosystems restored.

Ethanol and CBM fuel production will continue, as will bladeless wind-driven power and hydrogen production for airships and energy storage. These fuels and wind power generating technology will never be completely obsolete as long as there are airships, vegetation, livestock, and wind. Ethanol will continue to be a relevant fuel as long as small engines are required. SNG will be phased out and used only for energy storage once conversion from gas to electric has been completed globally and nuclear fusion becomes prevalent. SNG will then be replaced by CBM, which will be necessary as long as people continue to use cooktops with a gas flame. CBM will continue to be produced from RFC livestock urine and manure, and from sewage treatment and waste facilities. As CBM turbines are replaced by nuclear fusion, CBM will replace synthetic SNG for residential gas distribution.

Though nuclear fusion is the power of the future, only nuclear fission is available. It has the highest efficiency, is the least expensive per watt to construct, and it doesn't emit any greenhouse gases. Out of all the non-fossil-fueled technologies producing power, it is the second-leading global electricity producer behind hydroelectric. But is the deadliest if a catastrophe should occur. [122]

There're two different types of nuclear power plants in operation today worldwide, those using a uranium-plutonium reactor and others using a thorium-uranium reactor. Although there are advantages and disadvantages to both types, the thorium-uranium is less expensive, thus a better short-term solution. The nuclear plants that are currently under construction should continue, but redesigned to use thorium-uranium reactors instead of uranium-plutonium. Nuclear plant engineers that are currently in the planning stages should consider redesigning their plants to use the recently developed SMR technology, which will be approved and available in a few years. It will lower construction cost considerably.

New nuclear plants that are already operational should continue as they are and older nuclear plants, with outdated technology, must be mandatorily retired, or replaced with SMRs, once B100 biodiesel-fueled turbine REM sites are operational. All nuclear fission plants, regardless of which technology is used, will be replaced with nuclear fusion technology when it becomes available. [87][88][89]

A combined HCPVT system with a bladeless wind turbine system and a hydrogen electrolyzer with a storage tank for electricity storage, are a great combination for very remote regions where regular B100 delivery for turbine refueling would be difficult. Nuclear fusion will either replace or be added to these systems when the technology is available.

Solar PV cells should continue to be produced but only for use on solar-powered airplanes and HCPVT systems, or on outer space devices. Solar PV panel manufacturing should be discontinued until the efficiency has been increased, toxic waste reduced, and their recycling issue resolved.

Existing wind farms should replace their propeller-driven wind turbines with bladeless wind-driven turbines. Propeller turbines should be discontinued because of their danger to birds and for the enormous amounts of rare earth minerals required in manufacturing. Their cost of manufacturing versus their efficiency when they're compared to bladeless wind turbines, which have a 10% lower efficiency rating but a 51% lower manufacturing cost, makes them a less prudent choice for wind-driven power generation than bladeless turbines. Other advantages of bladeless systems include lower maintenance costs, no danger to wildlife, no noise, a lesser area footprint, and their lifespan is several decades longer than propeller-driven turbines. [182]

7

DISTRIBUTION AND TRANSPORTATION

Distribution and transportation systems within the RFC community will be designed for maximum efficiency. Pipelines, conveyors, and monorail transportation will be automated and have low maintenance and energy requirements. CBM-fueled turbines will power the water, sewage, and fuel pipeline pumps; conveyor hydraulic motors; and the elevated electric monorail that provides regularly scheduled service to the community's fisheries, municipal buildings, residential neighborhoods, businesses and retail shopping malls, and the food distribution and processing center.

A B100 biodiesel pipeline runs from the algae oil crop processing location to the RFC fuel depot and to the Maglev distribution terminal's fuel depot. And water pipelines will run throughout the RFC community, from the water distribution center and to algae ponds and grain crop locations outside the community.

Distribution conveyors are run throughout the community from the farms to the processing plants, then run to the food distribution and processing center for packaging and loading of each member's order for delivery. The conveyor systems for livestock feed go from the crops to processing and mixing then to the livestock locations. The farm waste conveyor runs to the ethanol fermentation processing location. The E100 ethanol fuel is then pumped through the pipelines from each processing location to the community fuel depot for storage and distribution.

A Maglev transit spur will be built from each RFC community to the closest Maglev distribution terminal. In the RFC locations that have their 6,000,000 members within a close delivery distance, the transit spur will not be used for their members' food distribution. It will only be used for commercial shipments, passengers, or specialty crops that are either received from another RFC or grown in the RFC for global distribution. A spur is only necessary for food distribution if a RFC has members a great distance away. The spur has a water pipeline and B100 biodiesel pipeline constructed within the spur's structure. The pipelines supply water and biodiesel to the Maglev transit distribution terminal for further distribution of water through the Maglev transit network to cities, and for B100 biodiesel storage at the distribution terminal's fuel depot for later distribution to REM sites. B100 biodiesel turbines will power the Maglev transit spur pipeline pumps.

RFC green communities in Africa and Asia may be constructed in regions where there are few existing roadways. Roads and bike paths will be built in these regions out of plastic recovered from landfills and oceans. The plastic is molded into large road brick sections designed for assembly on site by locking the tongue-and-groove-style bricks together. The recycled plastic roadway sections will have pipeline and electric cable space provided inside each plastic brick, under the road surface, for easy installation and maintenance. This technique will also be used to build the one-acre algae pond containment pools. Recycled plastic has been used frequently over the years for a variety of products and building materials. Over 96 trillion tons of mixed plastic waste is discarded every year and converting it will be the utmost priority. See Section 6: Electric Power and Fuel, and Section 20: Environmental Cleanup. [62]

$22,398,417,500 – Distribution conveyors, pipelines, monorail, roads, Maglev transit spur.

RFC trucks will pull 2,500 fifty-seat-capacity passenger trailers from the community's residential fisheries and neighborhoods to various worksites daily. Providing transportation to the farm and ranch work locations not serviced by the monorail. The trailers are dropped off at each worksite for the duration of the workday. Worksite locations change everyday, making these trailers the most efficient mass transportation method. Each trailer has two inserts that can be pulled out when parked, extending the inside width from 8 feet to 24 feet, increasing the dining capacity to fifty for lunch breaks. There will be over 125,000 RFC owner-workers and laborers transported every workday to various worksite locations within the RFC.

$50,000,000 – 2,500 passenger trailers at $20,000 each.

RFC green communities own all of the trucks stationed inside and outside each RFC community. And RFC Direct maintains the trucks outside each community and pays all of the truck drivers out of Maglev transit network commercial shipping revenue. RFC Direct also manages the Maglev transit network and schedules the RFC food deliveries and all of the commercial shipments outside each community's boundaries. See Section 16: Conglomerate.

RFC trucks within each green community will be used for various services including trailer transportation, hauling heavy machinery, and the community's residential food deliveries from the distribution and processing center.

Food orders will be either trucked outside each community to members' locations near the RFC, or will be shipped by flatcar from the distribution and processing center to the Maglev distribution terminal, where they are unloaded from flatcars and driven to the RFC members' regions. RFC Direct will dispatch all of the delivery trucks outside of the communities from each truck's designated Maglev distribution terminal.

The truck fleet operating outside each community is stationed at the Maglev distribution terminals and used for the RFC food deliveries between the hours of 10:00 PM and 6:00 AM. Then these trucks will be used for RFC Direct commercial deliveries between the hours 6:00 AM and 7:00 PM.

All trucks will be equipped for both delivery and maintenance service. Each truck driver will have the ability to repair any micro-grid that has a power outage, for faster service, if a power outage does occur in a delivery driver's neighborhood. Every delivery driver will be skilled in both truck driving and REM electrical repair and maintenance service. Electric power and repair service trucks outside RFC green communities are dispatched within a two-hour period.

Within each RFC community the maintenance and repair service trucks are dispatched within two hours and includes electrical, plumbing, and carpentry repair service. Every RFC driver in each community will be trained for any level of maintenance or repair required for residences, farms, ranches, factories, and buildings within the community.

Trucks outside the community:
$1,403,050,000 – 28,061 delivery-maintenance trucks at $50,000 each.
 1,403,050,000 – 28,061 commercial delivery trucks at $50,000 each.

Trucks inside the community:
$453,000,000 – 9,060 delivery-maintenance RFC trucks at $50,000 each.

For members living outside the RFC community, food orders will be delivered for pick-up close to each member's residence at REM sites in residential neighborhoods. Each refrigerator container has 192 lockers. When a refrigerator is loaded onto the Maglev flatcar it is plugged into the flatcar's power. When the refrigerator is unloaded at the distribution terminal onto a delivery truck, the refrigerator is plugged into the truck's power. And when the truck unloads the refrigerator at the REM site, the refrigerator is plugged into the power turbine supplying electricity to the neighborhood's micro-grid. The refrigerator does not require a power supply.

Each truck will deliver a tanker trailer filled with 2,000 gallons of B100 biodiesel when the refrigerator container is delivered. Simultaneous efficient delivery. After unloading the refrigerator, the empty refrigerator is then loaded onto the delivery truck and returned to the distribution terminal or to the RFC, along with the empty fuel tank trailer. The empty refrigerator is delivered to the RFC distribution and processing center for reloading and the fuel tank trailer is returned and refilled at the fuel depot for the next delivery day. There will be a combined total of 56,122 fuel tank trailers and refrigerator containers picked up and delivered to 28,061 REM sites. The refrigerator containers will be delivered every day. But the fuel tank trailers are only delivered when fuel is low at the REM site. Fuel consumption at each REM site varies depending on the total number of turbines at each site and the combined electricity usage of RFC members on each REM.

Every day, 3,189 refrigerators will be delivered to the residential neighborhoods and fisheries in each RFC community and 3,189 empties, from the previous day, will be picked up. There will be a total of 62,500 refrigerator containers used inside and outside each community.

RFC refrigerators will use ammonia (NH_3) or hydrofluoroolefin (HFO) as a refrigerant for cooling instead of hydrofluorocarbon (HFC), which is currently used. 80% of the world's HFC usage is for refrigerators and air conditioning units. It has a 14-year atmospheric lifespan and a global warming potential (GWP) that is up to 5,000 times greater than CO_2. HFC replaced chlorofluorocarbon (CFC) when it was phased out in the Montreal Protocol's 1987 treaty. CFC is banned for destruction to earth's ozone layer, high GWP, and 35 to 150 year atmospheric lifespan. In 2016, the treaty was amended to compel HFC reduction phase out in

2019. Its replacement is HFO, which is a greenhouse gas, but with a shorter 11-day atmospheric lifespan. Ammonia is flammable and can be hazardous when released into the atmosphere in high concentrations, but it is not a greenhouse gas. The type of refrigeration units used by each RFC community will determine the type of refrigerant required. [134] [142][143][145]

$1,122,440,000 – 56,122 B100 2,000-gallon fuel trailers at $20,000 each.
  1,250,000,000 – 62,500 refrigerator containers at $20,000 each.

    In special circumstances, an individual residential refrigerator locker will be offered instead of a designated pick-up refrigerator location. Individual home delivery refrigerator lockers, installed in residential neighborhoods, are powered by local REM power that is supplied to the member's residence. The refrigerators turn on automatically when a delivery is scheduled and off after the refrigerator has been emptied. When the delivery truck arrives, the driver removes the insert from the truck's refrigerator container and inserts it into the refrigerator locker outside the home. These refrigerator lockers could be configured and installed in either a multi-unit building or in a single-family residence.

    Any RFC member can apply for the neighborhood's installation of individual residential refrigerators. This convenience will be a separate one-time charge for each refrigerator and its installation, and a monthly fee for the additional delivery driver required to provide this service.

8
RFC GREEN COMMUNITY TOTAL COST

This is the total RFC cost estimate for all of the essential components necessary for construction and operation. The expenses are based on costs in the United States if each RFC community was constructed on undeveloped land. These costs will come down considerably in many of the world's regional locations due to currency exchange rates, lesser land value; larger volume discounts for equipment, vehicles, materials, etc.; and lower wages for construction labor.

$ 3,744,000,000 – 1,248,000-acre land purchase.
  3,000,000,000 – Ranch and farm budgeted start-up expenses.
 12,524,400,000 – 300 ten-acre concrete fisheries with condominiums.
  3,300,000,000 – $12,500 floor-finish x 264,000 floors in 300 fisheries.
  3,000,000,000 – Cleaning and processing equipment for 300 fisheries.
  7,893,760,000 – 24 plants: $328,906,666 each.
    225,000,000 – 2,250 miles of water pipelines at $100,000 per mile.
 11,287,440,000 – 125,416 three-bedroom houses at $90,000 each.
  1,254,160,000 – Farm and ranch buildings.
    975,000,000 – Community building prefabrication construction.
  5,625,000,000 – Five hospitals at a cost of $1,125,000,000 each.
    701,550,000 – 14,031 domes at $50,000 each.
  1,998,450,000 – Thirty factories at $66,615,000 each.
    100,000,000 – Industrial fabrication and 3D printing plant.
    100,000,000 – Fuel depot building, storage tanks, pipelines, equip.
    120,000,000 – E100 ethanol fermentation production equipment.
  1,000,000,000 – B100 algae oil refineries with pipeline to fuel depot.
  2,652,000,000 – 442 at $6,000,000 each anaerobic digester, compressor.
    442,000,000 – 442 at $1,000,000 each autoclave, mixer, ingredient, etc.
    221,000,000 – 442 at $500,000 each automated grass mower, rake, etc.
    442,000,000 – 442 at $1,000,000 each automated feedlot conveyor, etc.
      4,600,000 – Organic ingredient cultivation, harvesting, mixing, etc.
     76,800,000 – 64 1MW CBM-fueled turbines at $1,200,000 each.
    327,600,000 – 1,365 200kW CBM-fueled turbines. 273MW x $1.20/W
    120,000,000 – Hydrogen production equipment.

320,000,000 – Sewage treatment plants and waste management.
22,398,417,500 – Distribution conveyors, pipelines, monorail, spur.
50,000,000 – 2,500 passenger trailers at $20,000 each.
1,403,050,000 – 28,061 delivery-maintenance trucks at $50,000 each.
1,403,050,000 – 28,061 commercial delivery trucks at $50,000 each.
453,000,000 – 9,060 delivery-maintenance trucks at $50,000 each.
1,122,440,000 – 56,122 B100 2000-gallon fuel trailers at $20,000 each.
  1,250,000,000 – 62,500 refrigerator containers at $20,000 each.
$89,534,717,500 – RFC community cost without power to desalination
plants, RFC factories, or fisheries. Electricity will be
supplied from global power REM sites. See Section 12:
Global Cost and Section 13: Global Funding.

9

MAGNETIC LEVITATION TRANSIT

Unlike today's trains that have multiple cars pulled by an engine, the RFC magnetic levitation (Maglev) transit network will consist of 60-ft.-long self-propelled flatcars. Each flatcar will carry two 24-ft. containers or one 48-foot container. RFC containers are high cube, but all standard or high-cube containers from other freight companies can be shipped. Two containers between 20 and 24-ft. or one container between 40 and 48-ft. in length can be loaded onto any RFC flatcar. [138]

Each Maglev flatcar is equipped with a 200kW B100 biodiesel-fueled turbine generating enough power to travel at speeds reaching above 500 mph and lifting a total load capacity of 50 tons. These self-powered flatcars are fueled by biodiesel produced in RFC algae ponds. They will be A.I. computerized with destination software programmed to allow them to be directed to the correct distribution terminal without a human operator on board. Equipped with the latest safety equipment, pioneered by automobile manufacturers, allows each flatcar to sense any flatcar directly ahead. Flatcar accidents during transport at speeds above 500 mph, or at slower speeds when approaching, or within, a distribution terminal, will not be possible. A 200kW backup system is on each flatcar.

The Maglev transit network will operate 24 hours a day, 365 days per year. As many as 22,500,000 flatcars per day, loaded with commercial shipments or passengers, can run to distribution terminals located on six continents. This is in addition to the food shipments that are transported by truck to members' locations directly from each RFC community. The containers are offloaded at the distribution terminals onto RFC delivery trucks then driven to their final destinations.

TRANSIT CONSTRUCTION

Maglev transit track structures will be constructed using precast building techniques previously used for building bridges. This building technique was used in the United States for the construction of the bridges for the State of Florida's road connection to Key West. The tracks will be completely enclosed and airtight, with atmosphere removed to allow

Maglev flatcars to run unimpeded above 500 mph. The enclosed tracks eliminate any danger of anything getting onto the tracks and causing accidents when flatcars are running at these high speeds.

There will be 375,000 miles of four-track-wide structures built by 14,048 crews on six continents. Each of the 2,100-worker crews will build a 26.7-mile section over a seven-year period. Two years for the surveying and construction preparation and five years for the construction. The average cost per mile is $100 million. This includes bridges and tunnels. The water pipelines, run from the ocean water desalination plants, and the B100 biodiesel pipelines, run from the algae oil processing plants, will be built beneath the Maglev transit tracks and are an integral part of the transit track structures.

DISTRIBUTION TERMINALS

There are 120 major Maglev distribution terminals. There are thirty in North America, twenty in South America, thirty in Europe and Africa, and forty in Asia and Australia. Australia has four smaller terminals that are equal to one terminal. RFC green communities located there will serve as satellite distribution terminals to cover the entire continent. Each of these distribution terminals is positioned so they have a maximum radius of 250 miles from RFC members' regions for RFC deliveries by truck. Each major distribution terminal is approximately 35,000 acres. They will all be constructed simultaneously during transit construction. Surveying and construction preparation will be conducted over the first two years followed by five years of construction. A satellite distribution terminal will be constructed in each RFC community location and connected to their assigned distribution terminal by a two-track transit spur linking the RFC to the terminal. Each RFC spur's construction cost is included in each distribution terminal's $5 billion construction budget.

The Maglev transit and distribution terminal construction jobs and the increased hiring by the companies supplying essential parts and equipment will significantly boost the world's economy. And the Maglev transit network will ship containers globally faster than any other current form of transportation without emitting any greenhouse gases into the atmosphere. One massive container ship equals approximately 50,000,000 vehicles worth of greenhouse gas emissions. [71]

Maglev distribution terminals will have a fuel depot with a diesel pyrolysis processing plant for conversion of discarded plastic to diesel fuel. Stored at the facility, this diesel fuel will be distributed to non-RFC owned trucks, ships and machinery. The fuel depot will also have a B100 biodiesel distribution pipeline run to storage tanks for refilling REM fuel trailers and for refueling Maglev flatcars and RFC delivery trucks. See Section 6: Electric Power and Fuel.

The distribution terminals that are located on the coasts of each continent will have processing centers that convert discarded plastic and manufacture it either into road sections or into algae pond containment pools. These factories will be constructed in terminal locations where the largest accumulation of discarded plastic waste is delivered, exceeding the quantity that will be required for conversion to diesel. See Section 7: Distribution and Transportation. [62][67]

MAGLEV TRANSIT NETWORKS

There are four separate Maglev transit networks on six continents.

North America

Countries serviced: United States, Canada, Mexico

$ 7,341,846,835,000 – 82 RFC green communities.
   7,500,000,000,000 – 75,000 miles of 4 tracks X $100,000,000/mile.
     150,000,000,000 – Thirty distribution terminals at $5 billion each.
     600,000,000,000 – Maglev flatcar power: 2,500,000 – 200kW turbines.
     <u>250,000,000,000</u> – 2,500,000 Maglev flatcars at $100,000 each.
                     Flatcar cost includes containers with inserts.
$15,841,846,835,000 – 82 RFC communities with Maglev transit.

South America

Regions serviced: South America, Central America

```
$  7,520,916,270,000 – 84 RFC green communities.
   7,500,000,000,000 – 75,000 miles of 4 tracks X $100,000,000/mile.
     100,000,000,000 – Twenty distribution terminals at $5 billion each.
     600,000,000,000 – Maglev flatcar power: 2,500,000 – 200kW turbines.
     250,000,000,000 – 2,500,000 Maglev flatcars at $100,000 each.
                       Flatcar cost includes containers with inserts.
 $15,970,916,270,000 – 84 RFC communities with Maglev transit.
```

Europe and Africa

Regions serviced: Europe, Africa, Middle East, West and Central Asia

```
$32,680,171,887,500 – 365 RFC green communities.
 15,000,000,000,000 – 150,000 miles of 4 tracks X $100,000,000/mile.
    200,000,000,000 – Forty distribution terminals at $ 5 billion each.
  1,800,000,000,000 – Maglev flatcar power: 7,500,000 – 200kW turbines.
    750,000,000,000 – 7,500,000 Maglev flatcars at $100,000 each.
                      Flatcar cost includes containers with inserts.
$50,430,171,887,500 – 365 RFC communities with Maglev transit.
```

Asia and Australia

Regions serviced: India-Pakistan, East Asia, Southeast Asia, Oceania

```
$56,854,545,612,500 – 635 RFC green communities.
  7,500,000,000,000 – 75,000 miles of 4 tracks X $100,000,000/mile.
    150,000,000,000 – Thirty distribution terminals at $5 billion each.
  2,400,000,000,000 – Maglev flatcar power: 10,000,000 – 200kW turbines.
  1,000,000,000,000 – 10,000,000 Maglev flatcars at $100,000 each.
                      Flatcar cost includes containers with inserts.
$67,904,545,612,500 – 635 RFC communities with Maglev transit.
```

Each of the networks will service its RFC members from the RFC communities located within their network's regions.

Terminals located near a coast will accommodate airship landing for freight pick-up and delivery and passenger transit service. Airships service islands and other isolated regions. See Section 10: Rigid Airships.

The RFC Direct's Maglev transit network, distribution terminals, and algae acreage are global expenses. See Section 12: Global Cost, and Section 13: Global Funding.

$150,147,480,605,000 – 1,166 RFC communities with 375,000 miles of Maglev tracks and 120 distribution terminals.

6,334,206,710,000 – Maglev B100 fuel: 633,420,671 acres x $10,000 per algae pond containment pool.

152,157,041,000 – RFCs, transit routes, and distribution terminals surveying and preparation.

PASSENGERS

The 24-ft. passenger-inserts within the high-cube containers are basically private first-class compartments that can have up to six passengers each to accommodate people and their pets traveling together. Compartments are windowless, but have a video display, inside each compartment where windows would be, showing the outside terrain during travel. [149]

Advantages of six-passenger compartments per container:

1) You won't be exposed to airborne bacteria or contagious viruses from unknown passengers.
2) You can travel with up to six people per compartment plus animals.
3) You supply your own drinks and food.
4) You will be able to do anything: walk around, exercise, sleep, watch TV, use your computer or cell phone and the internet.
5) You can go to the restroom without having to wait in line.
6) You will not have to sit next to an obnoxious person or experience another passenger's child sitting behind you continually kicking the back of your seat.

7) It is less money to travel by passenger compartment than airline first-class or business class. If your family or friends travel with you, then the price of the ticket could be less than economy class seating for first-class travel.

8) People will be able to reserve a travel compartment online. Door-to-door pick-up and delivery service will be available. There won't be any waiting in a terminal for boarding. When passengers are picked up, or arrive at the terminal, they check-in and enter their compartment. After boarding, the compartment will be transported to the flatcar's location and loaded. It departs immediately. There isn't any regularly scheduled departure time.

UNDERGROUND TUNNELS

In order to connect the four Maglev transit networks and form a single intercontinental network, three underground tunnels will need to be built. This is challenging, but possible using the latest techniques for tunnel excavation and construction. The three tunnels will be under the Strait of Gibraltar between Gibraltar and Morocco; under the Bab-el-Mandeb strait between Djibouti and Yemen; and the longest and most challenging, under the Bering Strait between Alaska and Siberia. These continental links had been proposed before in the previous century, but were abandoned for various reasons. Mostly concerning underground tunneling technology and cost.

There are two options for the Strait of Gibraltar and the Bab-el-Mandeb strait tunnels. The first is tunneling underground. The second is building an underwater aboveground tunnel placed on the sea floor. The total estimated construction time for either option is no more than seven years. Two years for surveying and excavation preparation and five years for tunnel construction. The Bering Strait underground tunnel was first proposed in the late 1990s for an oil pipeline and has been approved. Preparations for construction have already begun. The estimated time for construction is approximately twelve years.

$150,000,000,000 – Transit tunnels: Strait of Gibraltar, Bering Strait, and Bab-el-Mandeb.

10
RIGID AIRSHIPS

Once the three transit tunnels under the Strait of Gibraltar, Bering Strait, and Bab-el-Mandeb strait are completed the global Maglev transit will be connected to North and South America, Asia, Europe and Africa. The island nations that survive the rising ocean levels will be serviced by airships. RFC Direct will fly airships with million pound load capacities to island nations throughout the world delivering commercial freight, passengers, and food and water shipments to RFC members. Though slower than airplanes, the cargo capacity is greater and they use less fuel. Airships have much less cargo capacity than ships, carrying a maximum of fifty 24-ft. containers, but are faster and eliminate the possibility of losing cargo in rough seas. See Section 16: Conglomerate.

Rising sea levels and the ever-increasing frequency of hurricanes and storms, from the raised ocean temperatures caused by global warming, make airships ideal freight transportation. They are able to fly around storms, giving them a great advantage over ships and airplanes due to their range of travel without having to refuel. Airships will use electric engines and their power turbines do not emit any greenhouse gases into the atmosphere.

Airships have removable artificial trees attached underneath their envelope. Protruding from their cargo hold during flight, they capture CO_2 from the air. When the airships dock, the trees are removed and soaked in water to release the CO_2, and then they are reattached to the airship. [114]

Two airships will be based in each RFC green community and an additional global fleet of 668 service islands and roadless rural regions on all continents. All 3,000 airships will use hydrogen for lift, supplied by each RFC from their fuel depot or from each ocean water desalination plant's hydrogen production facility. See Section 5: Houses, Buildings, Factories.

Hydrogen has been denied approval for use in airships for decades by the United States. It has been misrepresented as a dangerous fuel since the Hindenburg airship catastrophe in the evening on 6 May 1937. It is not more dangerous than other fuels. It is less flammable than gasoline. Gasoline can burst into flames at 442° Fahrenheit. Hydrogen gas won't ignite until 932°F. During the investigation of the disaster it was verified that the Hindenburg's outer skin had been doped with an extremely flammable aluminum/iron oxide, which caught fire when it was struck by lightning, well before the hydrogen ignited. Governments must approve hydrogen airships for flight globally. [162]

$450,000,000,000 – 3,000 airships at $150,000,000 each.

Airship distribution terminals will be combined with the Maglev terminals located on the coasts of each continent for freight pick-up and delivery. In addition to freight distribution, they are used for emergency transport to hospitals and for disaster relief in any devastated region. If a natural disaster does occur; such as an earthquake, hurricane, tornado or tsunami; an airship can respond faster than any other transportation, delivering food, water, blankets, temporary shelters, power generators, and transportable emergency medical facilities, with airlift drones, to the disaster affected areas.

Every Risk Reconnaissance Group (RRG) team will be trained to respond to disasters, supplying emergency medical relief to the injured and providing protection from looters. RRG teams work together and coordinate with other RRG teams and the local authorities to eliminate duplication of areas that are already covered in affected regions. This will provide the absolute best possible relief service to the most residents. See Section 16: Conglomerate.

11
COMMUNICATION

The RFC green community network relies on its communication service. Without it this global solution will not work. RFC members must be able to place their orders without any communication interference. There are many regions of the world where internet or cell phone communication is not available, or if available, has poor service.

Equality amongst people throughout the civilized world includes the highest-level of communication services, not just food, water, power, and healthcare. The ultimate goal of the RFC green community's global solution does not end with only the basic requirements offered. A true egalitarian society elevates everyone in the world to the highest possible level, which is to say Japan, Singapore, South Korea, European Union, Australia, Canada, and United States. These countries use the highest-level of communication technology. To end famine and extreme poverty the world must elevate the poor and developing countries to the highest level too. This egalitarian philosophy includes communication.

RFC communication is a great deal more than an infrastructure for the internet and cell phone network. There will be thousands of hours of development and programming that goes into the network database to allow members to order online. Over seven billion members will have personal data stored on the RFC network servers, including demographic information, ordering and delivery history, retirement account, Maglev transit travel information, etc. Data is encrypted and kept confidential. The RFC ordering website and cell phone app advertisers will never have access to any member's personal information. For members to access their own personal account data requires a fingerprint, a retinal scan, membership login access code and password.

The following software and database programming tasks and the equipment for the network are included in the communication expenses:

1) Global datacenters and the communication network for internet and cell phones with secured servers, regulated electric power supply, and system infrastructure.
2) RFC members' ordering database, website, and cell phone app.

3) Global Survival Fund donation database, website, and cell phone app.
4) RFC Direct passenger travel and cargo ordering database, website, and cell phone app.
5) Flatcar destination software programming.
6) Conglomerate websites Risk Reconnaissance Group, Green WALET, and Zone Interactive. See Section 16: Conglomerate.

Every RFC community will use the communication network built into their REM Smart technology infrastructure to guarantee service to every resident. This will include every REM site outside the community. All cell phone and internet services will be included with each RFC membership if service is not available in a member's region. This will assure communication is everywhere for members. The communication network infrastructure is a global expense and not a RFC expense. See Section 12: Global Cost, and Section 13: Global Funding.

$3,000,000,000,000 – Programming, datacenters, and communications network.

12
GLOBAL COST

Global cost is the estimated total of RFC communities' expenses and outside the communities' expenses that are not part of RFC community construction costs, but are required to accomplish the RFC global seven-year goals that are stated in this plan. All of the cost estimates are based on construction in the United States. These high costs will come down considerably in many countries due to the lower currency exchange rates, lesser land values; larger volume discounts for the equipment, vehicles, materials, etc.; and their lower wages for construction labor.

Every RFC community has a scalable design to allow it to expand its capacity above their 6,000,000 members whenever new members are added, without membership fee increases or an additional member cost. Expanded maximum capacity is calculated at being double its current 6,000,000-member capacity. With the capability of scalable expansion of their production, there shouldn't be any need to build additional RFC green communities until 2110, based on the predictions at the current rate of global procreation. See Section 13: Global Funding.

The RFC green community plan's total cost includes all 1,166 RFC communities and their Maglev transit distribution network. The Maglev transit total cost includes the enclosed tracks with all pipelines constructed within the elevated structures, vacuums for removing the atmosphere from the enclosed tracks, the distribution terminals with fuel depots, a two-track transit spur to each RFC green community; flatcars, containers, inserts, B100-fueled power turbines; the global algae pond acreage for B100 biodiesel production, transit route and RFC surveying, and the construction of the three underground Maglev transit tunnels. See Section 9: Magnetic Levitation Transit.

$150,147,480,605,000 – 1,166 RFC communities with 375,000 miles of Maglev tracks and 120 distribution terminals.

6,334,206,710,000 – Maglev B100 fuel: 633,420,671 acres x $10,000 per algae pond containment pool.

152,157,041,000 – RFC and transit route surveying and preparation.

150,000,000,000 – Transit tunnels: Strait of Gibraltar, Bering Strait, and Bab-el-Mandeb.

REM sites are scalable to accommodate the amount of electricity required during peak demand hours by each REM site's RFC members. A shared infrastructure design will be scalable to add or remove power-generating or communication devices without shutting down the site. REM power turbines will use B100 biodiesel fuel, produced in one-acre algae ponds, until nuclear fusion becomes available. Each site will be a drop-off location for RFC refrigerator container delivery. See Section 6: Electric Power and Fuel, and Section 7: Distribution and Transportation.

$15,360,000,000,000 – RFC electric power: 12,800,000 – 1MW operational REM sites.

9,775,261,540,000 – RFC B100 fuel: 977,526,154 acres x $10,000 per algae pond containment pool.

4,000,000,000,000 – REM sites and electrical and communication infrastructure with built-in Smart technology.

RFC communication services includes cell phone and internet. See Section 11: Communication.

$3,000,000,000,000 – Programming, datacenters, and communications network.

Water pipelines will be built into Maglev transit track structures, but many cities and towns will not be close to the Maglev tracks or the terminals. These global funds pay for water lines not integrated into the transit track structures. See Section 4: Water Requirements.

$700,000,000,000 – Potable water global distribution pipelines.

The grain producing acreage will not be located within the RFC green communities. Grain crops will be grown in particular regions that are good for grain cultivation, but may not be as desirable for other RFC crops. This acreage is a global expense. See Section 2: Land Requirements.

$1,080,000,000,000 – 360,000,000 grain crop acreage at $3,000 per acre.

RFC owner-workers begin their newly assigned jobs and living in their community at the end of the sixth year of RFC green community construction. Global funds will pay their salaries until RFC communities are operational. See Section 21: RFC Forty-Year Timeline.

$1,100,000,000,000 – Salaries for Green WALET and RRG personnel,
      and RFC owner-workers prior to 8th yr.

The restoration of the environment to its previous condition is expensive, but it is an essential component of this comprehensive global RFC green community plan. See Section 20: Environmental Cleanup.

$1,000,000,000,000 – Ocean cleanup: aquaculture, wetlands, and coral
      reef restoration.
  426,108,080,000 – Ranch Farm soil contamination cleanup, fertilizing,
      planting.

Wildlife protection is pertinent in every country, but especially vital in countries where a significant number of endangered and almost endangered animals continue to be the targets of poachers, and illegal hunters and trappers. This includes protecting the marine reserves from illegal fishermen in ocean regions. See Section 2: Land Requirements, and Section 19: Animal Rights and Wildlife Protection.

$400,000,000,000 – Wildlife preservation security, defense operation and
      equipment.

Airships will be a shipping improvement for isolated regions and provide faster aid to victims in disaster-affected regions than any other transportation. See Section 10: Rigid Airships.

$450,000,000,000 – 3,000 airships at $150,000,000 each.

The total global cost is an enormous amount. Inconceivable, until it's compared to the costs of a mass annihilation from the cataclysmic consequences of global warming. Though the earth will continue on, the human race may not. And although governments were quick to bail out banks in 2008 with, "Too big to fail," as a rallying cry, they have been less enthusiastic to come to our planet's aid. In this celestial neighborhood there are no other habitable planets on the market for a less expensive price. It is anticipated that the global cost will be a significantly lower amount than the estimated total. But whatever the final cost is, it will be a bargain compared to the exorbitant penalty the world will pay if nothing is done immediately to prevent the imminent environmental catastrophe that is rapidly descending upon us. See Section 1: Project Concept and Design, and Section 13: Global Funding.

$194,075,213,976,000 – Total global cost.

13
GLOBAL FUNDING

While contemplating this massive RFC green community challenge, the task of paying for this global solution was continually on my mind. The last thing I wanted to do was think of a solution without a way of paying for it. I am not a politician, an elected or appointed government official, that continually comes up with great ideas without a way of paying for them, leaving taxpayers to foot the bill. That is an unacceptable policy.

Consequently, it is important to me, my family, and the other 7.6 billion people that are occupying this planet that I literally represent in this endeavor, to present a solution that actually has a chance of working for the entire planet. Instead of a plan suitable for one nation's citizens and detrimental to another, which seems to be the way it has worked for the last several centuries. So this plan with funding and payback method had to work for everyone. [72]

I designed this plan as comprehensive as possible to be fair for everyone including the other species on the planet that, without a voice, are vulnerable to the decisions of the people we have placed in charge. This is as close as I could come to accomplishing an egalitarian design. This is how to fund this global solution and repay the funding.

BOND FUNDING

Beginning in year 2020 through 2021, RFC Direct will issue 354,000,000 twenty-year bonds for $14,000 each at a 4% interest compounded return rate totaling $4,956,000,000,000, to pay expenses incurred during the first two years of this plan. In year 2022, RFC Direct will issue an additional 13,508,515,284 twenty-year 4% bonds for $14,000 each. See Section 16: Conglomerate and Section 21: RFC Forty-Year Timeline.

$194,075,213,976,000 – Total funding from twenty-year bonds.

This is a massive amount of funding and before we continue on to the "Use of Proceeds," monetary comparisons are essential. The size of the entire world's gross domestic product is $50 trillion. The value of the world's stock and bond market is over $100 trillion. The global financial derivatives market is $700 trillion, which is over three and a half times more than the maximum global cost of the RFC green community plan and Maglev transit network. In October 2018 the stock and bond market lost $5 trillion. And during the six months from May 2018 thru October, China's stock market lost $3 trillion. Are we in the beginning of what will be the greatest financial crisis in history?

The global financial boom the world has experienced for the past decade was triggered by the bank bailouts in 2008, charging them low 0.00 to 0.25% interest rates until 2015, and allowing them to create the monetary conditions that led us to the economic boom. Which is now leading us directly to economic bust and currency collapse. Economic turmoil and currency collapse are already being seen in Argentina, India, and Turkey. Financial conflicts almost always lead to military conflicts. It wasn't the enacted government programs after the stock market collapse in 1929 that prompted the world's emergence from the great depression. It was World War II economics financed by war bond sales. [156]

The RFC's mission achieves similar economic relief. But instead of a war, it curbs global warming, ends famine, improves healthcare, and saves the world.

Use of Proceeds

$150,147,480,605,000 – 1,166 RFC communities with 375,000 miles of Maglev tracks and 120 distribution terminals.

15,360,000,000,000 – RFC electric power: 12,800,000 – 1MW operational REM sites.

9,775,261,540,000 – RFC B100 fuel: 977,526,154 acres x $10,000 per algae pond containment pool.

6,334,206,710,000 – Maglev B100 fuel: 633,420,671 acres x $10,000 per algae pond containment pool.

4,000,000,000,000 – REM sites and electrical and communication infrastructure with built-in Smart technology.

3,000,000,000,000 – Programming, datacenters, communication network.

1,100,000,000,000 – Salaries for Green WALET and RRG personnel, and RFC owner-workers prior to 8th year.

1,080,000,000,000 – 360,000,000 grain crop acreage at $3,000 per acre.
1,000,000,000,000 – Ocean cleanup: aquaculture, wetlands, and coral reef restoration.
700,000,000,000 – Potable water global distribution pipelines.
450,000,000,000 – 3,000 airships at $150,000,000 each.
426,108,080,000 – Ranch Farm soil contamination cleanup, fertilizing, planting.
400,000,000,000 – Wildlife preservation security, defense operation and equipment.
152,157,041,000 – RFC and transit route surveying and preparation.
150,000,000,000 – Transit tunnels: Strait of Gibraltar, Bering Strait, and Bab-el-Mandeb.
$194,075,213,976,000 – Total global cost.

BOND MATURITY PAYOUT REVENUE

Twenty-year maturity payout on each $14,000 bond is $30,675.72.

13,862,515,284 bonds x $30,675.72 = $425,242,637,347,704

Advertising

4,481,436,086 – Global minimum ordering sessions per day. This includes food and water orders, Maglev passenger travel reservations, freight shipment scheduling, and healthcare appointment online scheduling.

$4.50 – 18 ads displayed during an order session at $.25 per advertisement.
$4.00 – bond maturity revenue.
$0.50 – websites, datacenter, and communications network operating and maintenance expenses.

365 days per year x 20 years = 7,300 days

4,481,436,086 ordering sessions x 7,300 days x $4.00 maturity revenue = $130,857,933,711,200 – Total availability for twenty-year bond payout on maturity dates.

Shipping

12,104,964 – 60-ft. flatcars per day shipping two 24-ft. or one 48-ft.
        container per flatcar.

24,209,927 – 24-ft. containers per day at an average rate of $3,800 each.
12,104,964 – 48-ft. containers per day at an average rate of $7,600 each.

$3,800 charge breakdown - $2,500 – bond maturity revenue.
                          1,300 – operation and maintenance.

$7,600 charge breakdown - $5,000 – bond maturity revenue.
                          2,600 – operation and maintenance.

365 days per year x 20 years = 7,300 days

24,209,927 – 24-ft containers per day x 7,300 days x $2,500 bond revenue =
$441,831,167,750,000 – Total availability for twenty-year bond payout on
        maturity dates.

      Airship commercial shipping and passenger transport to islands
and isolated rural roadless regions is included in the container per day
shipping total. Airships can carry fifty 24-ft. containers or twenty-five 48-
ft. containers or any combination of the two up to a one million pound
total load capacity. See Section 10: Rigid Airships.
      All the twenty-year bonds' sales revenue is used in the first seven
years. The bond maturity payout revenue from advertising, commercial
freight, and passenger transportation starts accruing at the beginning of
year eight and continuing through the end of year twenty-seven. Payouts
begin on maturity dates as the first bonds sold in year one mature in year
twenty. Bond maturity payouts will begin in year twenty and continue
through year twenty-seven.

BOND MATURITY PAYOUT EXCESS REVENUE

The bond maturity payout plan has been designed to exceed the payoff requirement. Using these revenue-generating methods guarantees both bond payoff and additional excess income. Ensuring that all the quotas and the revenue from advertising and shipping are met is the utmost priority. If everything proceeds according to the intended plan there will be $147,446,464,113,496 of excess funds after bond maturity payoff. This money will be used for several purposes.

In order to achieve success, the countries of the world will need to endorse this global solution. If government backed financing is not required and government involvement is not essential, except for their ratification of the ten requirements listed in "Section 1: Project Concept and Design," then to obtain a consensus another incentive for endorsing this plan must be presented. Without each government's cooperation this plan can succeed, but it will have much less opposition with a unanimous worldwide endorsement. Currently, the combined external national debt of every country exceeds $82 trillion. But, if governments can stabilize their debt, or reduce it with the increased revenue from the improved global economy guaranteed by this plan, then the first expenditure of the excess funds will be to payoff these external national debts. [3][4]

This fund guarantee does not give countries consent to continue to spend excessively. To prevent abuse, this guarantee has a stipulation attached. Each country will have to balance their budget to prevent any additional debt increase from occurring. Each debt must have a cap. If world governments agree to this provision, then governments and their countries' citizens have an accord.

The excess funds will be allocated to the following objectives:

1) Payoff the external national debts of every country accepting this RFC global solution.
2) Funding for college education tuitions for RFC owners' children that have graduated high school.
3) Funding for specific mosquito species eradication and vaccines for mosquito-borne diseases. [105]
4) Funding for medical research and technology, purchasing equipment for RFC hospitals and medical centers, and training for medical staff.

5) Funding for R&D of innovative methods of energy generation and transportation. Improving terrestrial distribution and advanced outer space transportation methods.
6) Funding for off-planet research and colonization of the moon and Mars. Whether RFC members reside on Earth or not they will require food, water, electricity, and healthcare.
7) Decommissioning of nuclear fission power plants and the removal of all radioactive nuclear waste from the current onsite storage locations. Waste is shipped by airship to a launch site and then transported to the moon, where a storage facility will be constructed.

Once nuclear fusion is worldwide the decommissioning of fission reactors can begin. Storage facilities for high-level nuclear waste have previously been built, but currently, only the WIPP is open. There isn't any terrestrial location where radioactive waste will be guaranteed safe for thousands of years. It must be removed from the planet. The moon is the logical solution. [157]

It is crucial that the RFC green community plan expand to off-world colonization. At the current unabated rate of human procreation, we will have no other option but to colonize other worlds.

Populations continue to increase and compelling high schools to include procreation awareness and consequence courses, increasing the distribution of contraceptives, and promoting sexual abstinence globally, have been futile attempts to slow down procreation. Another solution, which doesn't involve contraceptives or sexual abstinence, would greatly slow procreation down. Eliminating unwanted pregnancy would impede over-population growth. If male children entering puberty were required to ejaculate a predetermined amount of sperm to store cryogenically for later artificial insemination into their future spouse or surrogate mother then afterwards get a vasectomy, it would reduce over-population growth.

There are five advantages to this vasectomy solution:

1) Couples could have sex without the worry of an accidental pregnancy.
2) Women would no longer be the gender responsible for birth control.
3) Pregnancy could be planned out prior to artificial insemination.
4) Abortion wouldn't be performed except in a life-threatening situation.
5) Backup sperm: if medical condition or accident causes male infertility.

This should be satisfactory to the pro-life advocates, the women's pro-choice advocates, and in the United States, nullify the controversy of their Supreme Court's ruling on legalization of abortions in Roe v. Wade.

If a male perceives he is not getting his spouse pregnant then he should have sex with her prior to insemination. This solution is better for women, eliminating the need for body altering birth control methods, and men are not affected in sexual performance. To eliminate accidental loss of sperm it will be stored in several different cryopreservation sperm bank locations. It can be released for planned insemination using a retina scan and fingerprint. This ensures sperm is safe and secure for future use.

According to one U.N. report: the global population reached an estimated 7.6 billion in 2017. It is predicted the global population will reach 8.5 billion by 2030, 9.7 billion by 2050, and 11.2 billion by 2100. With an ever-increasing global population, the earth's finite resources will not remain sustainable.

If vasectomies are unacceptable and voluntary birth control has been unsuccessful, what protocols should governments consider to curb our world's over-population growth? Shall it be a government-enforced limit on pregnancy (China's one-child policy was in effect for 35 years and had disastrous results), or global genocide? Such as intentional mass extermination that begins by limiting availability of a vaccine for deadly pathogenic bacteria once it has been released from thawing permafrost under melting ice sheets. Denying the vaccine to entire regions of human beings could reduce the global population by pandemic.

Or perhaps something a little less sinister such as a world war, which would basically be genocide triggered by a financial crisis and an economic collapse. This scenario could simultaneously mass exterminate and renew the economy. But the least offensive solution to our quandary is off-world colonization. There is nowhere else to go but up. [153][154]

However, an unintentional mass extermination may be lurking in our future. What global plan is there for celestial threats? We fortuitously dodged several celestial objects in recent years, but a direct hit could quickly eradicate most of the life on our planet. Although this may be a preferable fate when compared to slow obliteration by accelerated climate change, I believe there is enough time to prevent both possibilities.

Governments have failed miserably in their disgraceful attempts to resolve terrestrial threats. If governments can't adequately curb global warming, how do they intend to cope with an asteroid on a collision course with Earth? In their current climate change agreement, approved during the 2015 Paris conference, governments agreed to allocate $100 billion a year to spend on finding a solution to the global warming crisis, until 2025, when they rationalize yet another inadequate collective goal. Which will lead to further worldwide frustration and anger. Meanwhile, our hopeless leaders will bury their heads in the sand and ignore the fact that governments are not equipped to defend our planet from a celestial threat. The same inept way they previously ignored the veracity of climate change, but still fail to allocate enough funding to advocate a solution.

When NASA opened the Planetary Defense Coordination Office (PDCO) in 2016, with an annual budget of $150 million allocated to track Near-Earth Objects (NEO) to identify any potential threats; it affirmed the governments' stance. Similar to allocating funds that are trillions of dollars less than what is required to curb global warming, the PDCO is underfunded, which limits outer space observation to detect NEO threats from within our inner solar system and hinders the overall development of the technology necessary to prevent a NEO collision.

Though an estimated 95% of all NEOs have been mapped, 5% remain uncharted. And the likelihood of failing to detect a NEO traveling within the immense space surrounding our planet is a high probability. If on a collision course, the closer that the NEO gets to Earth without early-warning collision detection decreases global response time.

The PDCO's National Near-Earth Object Preparedness Strategy and Action Plan does confirm there isn't any existing NEO deflection technology developed to prevent a celestial collision. Rather than waste the $100 billion a year on inadequate solutions to global warming, they should allocate that money to fund the development of an early-warning detection system and deflection technology to defend our planet from celestial target practice. [1][80][81][102][103][130][174]

14
RETIREMENT BENEFITS

The RFC retirement program is an optional program, not mandatory. It is available for children 3 or younger after each RFC green community begins operation in year eight. Prior to turning 3 years old, children are covered under their parents' membership. The idea of the program is to allow parents to double their child's membership payments when they become a member at age 3, until age 25, when the young adult member takes over their payment responsibility and continues paying their dues until turning 50. When the member reaches 50, their annual membership payments cease, but their membership continues to be paid annually by the program for the rest of their life. It is a simple reliable retirement plan designed for every member so they won't have to worry about their RFC annual membership dues in their middle age through golden years, when they should be enjoying life.

When each double payment is received, half of it pays for the current membership dues and the other half is paid into the member's retirement account that is jointly owned by RFC Direct. The only way that funds for annual membership payments can be released from the account is through joint approval from the member and RFC Direct after the member turns 50. This will assure that funds are always there and members will not have to worry about funds being released without their explicit permission. Both retina scan and fingerprint are used to confirm each member's identity for the annual payment approval. See Section 16: Conglomerate.

However, when the member dies the remaining funds, if any, are automatically released to the Global Survival Fund as a donation for a destitute member that is not able to pay their annual membership fee. This is only random if the member dies unexpectedly. If the member is dying and wishes to release the remaining funds in the account upon death, the member can choose the destitute member through the Global Survival Fund to release the money to as a donation. The first choice is someone that has reached 97 years old and their account has run out. To continue the membership past 97 years, funds from accounts whose members are deceased add to the 97-year-old member's account until their death. The second choice is anyone, including a family member, that is having difficulty paying their dues, but it cannot be someone that is financially able to pay their annual membership dues.

This retirement plan has no restrictions on whether you work or not. Regardless of whether you continue to work after age 50, RFC Direct begins paying each person's annual membership dues starting at age 50 until their death. This plan will guarantee all people will have food, water, power, and healthcare for life.

Using the highest membership dues as an example, this is how the retirement program works: Each RFC member pays $4,200 per year for membership. For every newborn child whose parents are members the first 3 years are free. When a child reaches age 3 the parents pay $4,200 each for membership plus another $4,200 per year for the child's membership. Then an additional $4,200 per year is paid into the child's retirement fund. The parents pay the total of $8,400 per year per child until a child becomes a 25-year-old young adult. This is when the parent's payment responsibility ends. This gives the young adult the ability to get an education through a master's degree by age 25 without worrying about membership dues. At age 25, the young adult takes over and begins paying $8,400 per year. At age 50, the payments cease and the $4,200 annual membership dues are paid every year from the member's retirement account until the member's death. With better RFC healthcare provided in the membership, life expectancy will increase.

There will be a mandatory fee for membership additions that is separate from the membership retirement payments. Each RFC green community will begin with a minimum capacity of 6,000,000 members. The population of the world is increasing yearly and though the RFC communities are designed to expand their capacity, once that maximum capacity has been reached then new RFC communities will need to be built. When this happens an additional charge will be levied per newborn child of $27,741 for RFC new construction expenses. This fee can be paid either immediately or in twenty interest free installments of $1,387.05 per year, which can be further divided into monthly payments beginning when a child is born. A payoff can be made at any time.

Schedule of RFC retirement program payments

Age

0 to 3	Parents pay $1,387.05 per year for new RFC construction.
3 to 21	Parents pay $4,200 per year for membership plus $4,200 per year retirement fund and $1,387.05 per year for new RFC construction.
21 to 25	Parents pay $4,200 per year for membership plus $4,200 per year retirement fund.
25 to 50	The adult member pays their own $4,200 per year for membership plus $4,200 per year retirement fund.
50 to end	Annual membership dues are paid from the retirement fund.

Currency valuation in every country is a concern. My hope for the future is that every country will have the same currency valuation, eventually leading to a world currency. Until this happens each RFC green community operates under the currency valuations of the country that the community is in and the country where its members are located. This will not affect the retirement funds. If the membership dues increase because of the status of the country's economics, the previous payments into the retirement funds are unaffected. The previous payments do not have to be adjusted for the higher value of the currency.

Parents of children over the 3-year-old initiation age, and under the age of 25, can enroll their children into the retirement program. But the yearly payments missed, between the time the child turned 3 years old up to their current age at the time of enrollment, will have to be paid into the account prior to the program's initiation. If the membership dues have been adjusted since the child turned 3, because of currency valuation and the economics of the country where the member resides, then those missed payments will be calculated at the current membership rate at the time of the enrollment.

For older members that are over 25 and wish to enroll in the RFC retirement program, a RFC account will be initiated to deposit their extra payments for annual release when the member reaches retirement age. However, it is suggested that a savings or retirement account at a banking institution that pays interest may be more appropriate.

15
BUSINESS MEMBERSHIPS

Restaurants, bakeries, etc., and food-processing businesses are allowed to be members of the RFC green communities. There is a special rate for food-related businesses based on the quantities of food, water and power they use per month. Healthcare is not included in business memberships. Other types of businesses that are not food related are allowed to be members also, but their memberships are only for electricity and water. Their membership rates are calculated according to their electricity and water usage. All businesses are thoroughly inspected by Green WALET for their use and efficiency prior to a membership being granted. See Section 16: Conglomerate.

All restaurant and food-processing establishments must adhere to the RFC requirements for cooking and processing. Health and safety are a major concern and any cooking or processing of food must pass the strict RFC health and safety requirements prior to an establishment being granted a membership. Health departments located outside RFC green communities will partner with Green WALET on inspection procedures that are crucial to maintaining the RFC standards for food preparation.

RFC members that don't cook will be allowed to designate food delivery to an RFC-approved private meal preparer or restaurant that caters to those members that don't cook or don't have a place to prepare meals. This allows members to enjoy the benefits of home-style prepared food in their own home without having to go out to eat at restaurants or order restaurant takeaway. A list will be provided to members interested in this type of service. Members will pay meal preparers for their cooking services, but the ingredients for each prepared meal are already paid for by each individual's membership.

Local small markets that are RFC members will be able to stock basic produce. This is for those RFC members that have a quick need to get something they forgot to order or for those times, for instance, they accidently drop a tomato on the floor, it splatters, and they must go to the local market to retrieve another one. Payment will not be required to purchase it, but their RFC member's identification number will need to be presented at check out.

16
CONGLOMERATE

The conglomerate consists of each individual Ranch Farm Cooperative free-trade zone green community and five foundations that are a vital part of the RFC global solution.

RFC DIRECT

RFC Direct is the governing body and worldwide distributor for the RFC communities that will be constructed using RFC Direct issued bond sales revenue. It is a nonprofit foundation funded entirely by the commercial shipping revenue from the RFC Direct global transportation network of Maglev flatcars, trucks, and airships. It coordinates the delivery of food and water ordered from each RFC to their members' regions globally. RFC members pay a yearly fee that allows them to order food and water, produced by their RFC, 24 hours in advance. Orders are placed using the RFC Direct website or cell phone app. See Section 11: Communication, and Section 13: Global Funding.

Every year RFC members fill out an online form listing the types of food they want their RFC to produce. This is entered into the RFC Direct database that keeps the daily records of each member's ordering routine for selected foods and water amounts. Each RFC then produces according to the average frequency each member orders particular types of foods and water. Crops, livestock and fish are produced based on ordering habits.

Each membership includes electric power and healthcare. Power will be supplied directly to each member's residence from a REM site, operated by RFC Direct, outside the boundaries of the communities. Healthcare will include medical services from RFC hospitals and medical centers located both in and outside each community's boundaries. If a member moves to another region their membership and ordering record moves to another RFC.

RFC Direct will operate the global Maglev transit network and the airships, and operate and maintain the trucks for overnight deliveries, and operate REM sites outside each community. The orders are delivered directly to a member's residence or to a designated neighborhood REM pick-up location that is close to a member's residence, for pick-up from their RFC's refrigerator locker.

RFC Direct operates and maintains the global communications network. This service provides internet and cell phone communication to all RFC owners and members without service.

The RFC business memberships will have special calculated rates. Food-related businesses will be supplied food, water, and power. Other businesses will be supplied water and power only.

GLOBAL SURVIVAL FUND

Global Survival Fund is a nonprofit foundation that collects event and website donations for RFC Direct and distributes them to RFCs for destitute members that can't afford RFC memberships. Each donation in its entirety goes directly to the membership fee with no administration expense subtracted from the donation. The donation is routed to the RFC green community that is closest to the recipient of the donation.

Each donation will purchase a selected individual or family a yearly membership from the closest RFC to their region. This is a regular annual membership that supplies food, water, power and healthcare to the destitute member or family. Housing can also be supplied through other donations.

In order to uphold the strict, "no administration fee from any donation," policy, Zone Interactive website administrators, that maintain the website content and database, are paid by advertising revenue from the Global Survival Fund and RFC Direct websites and cell phone apps. The Global Survival Fund website is designed to collect donations and process them through the selected RFC green community's membership administration supervised by RFC Direct. Each RFC is already set up to process donations as they would any other paying membership. All RFCs can receive membership applications for donation assistance. The Zone Interactive administrators post them to the Global Survival Fund's website. This procedure assures that 100% of every donation goes for membership fees. See **ZONE INTERACTIVE**.

GREEN WALET (pronounced "wa lay")

Green WALET (**W**ater, **A**griculture, **L**ivestock, **E**nergy, **T**ransportation) is a nonprofit foundation entirely funded by Zone Interactive advertising revenue. It consists of a think tank, consultants, and inspectors that do research, then propose and implement new methods and technology to improve the efficiency and performance of each RFC green community.

This includes supervising RFC construction of hospitals, housing, community buildings, schools, food processing facilities, ocean water desalination plants, energy production, and the RFC community transit. They oversee RFC Direct's Maglev transit network, airship transportation services, and RFC regional and commercial delivery trucking.

Green WALET observes and inspects the food-processing plants, regenerative agriculture and organic crop rotation, livestock feedlots and slaughtering, fresh and saltwater fisheries, ocean aquaculture restoration, and wildlife refuge and sanctuary areas and migration routes adjacent to RFC green communities. Green WALET is also the governing body for Risk Reconnaissance Group.

Green WALET inspectors are able to accurately pinpoint any type of food contamination prior to shipment or, if shipped, accurately pinpoint the location the product came from by using RFC Direct computerized tracking methods for member deliveries from each RFC. This eliminates any confusion as to where the RFC product originated and prevents any food contamination epidemics from ever occurring. See Section 2: Land Requirements.

These consultants and inspectors are selected from universities, colleges, and the private sector, and from RFC ranchers and farmers. They are experienced in the latest organic regenerative agroecology techniques, livestock, aquaculture, and the required fields of engineering. They have no association with any company related to their field of consultation and inspection. There are no exceptions to this rule. If a conflict of interest is detected the decisions will always be under review from other Green WALET inspectors.

RISK RECONNAISSANCE GROUP

Risk Reconnaissance Group (RRG) is a nonprofit foundation consisting of multilingual retired military observation trained personnel that are vetted by Green WALET prior to approval for RFC green community placement. One hundred RRG operatives per RFC are owner-workers that live in their assigned RFC community. RRG operatives are governed by the Green WALET foundation. The following is a job description for RRG operatives:

- Land and sea reconnaissance and surveillance both inside and outside the RFC green community borders.
- Design emergency preparedness plans for the RFC community.
- Protect the RFC community during emergency and run the appropriate emergency preparedness plan.
- The RRG team is in charge of the community during an emergency and outrank everyone.
- Protect wildlife outside each RFC boundary from poachers, and illegal hunters and trappers in cooperation with local law enforcement.
- Minimum of 20 RRG remote control pilots fly solar airplane surveillance and drones outside a community's boundaries for wildlife protection.
- Protect RFC community perimeter against wildlife, natural disasters, trespassers, etc.
- RRG global emergency response teams transport supplies to a disaster-affected region.
- RRG operatives train RFC residents in observation security techniques.
- RRG teams operate during all public events at venues within the RFC community.
- All 100 RRG trained RFC owner-workers have a minimum of EMT medical training or above.
- Minimum of 15 RRG operatives have training in bomb and WMD defusing.
- Minimum of 20 RRG operatives are trained sharp shooters.

On-duty operatives wear their identification on shirtsleeves: **RRG**
RFC 0000

RFC-issued ID tags have their RRG designation along with the RFC community number.

The RRG website, www.riskreconnaissancegroup.org, and phone apps will post emergency information.

ZONE INTERACTIVE

Zone Interactive is a nonprofit foundation that produces Global Survival events, documentary films, and regional and international meetings and conferences. It is the administrator in charge of maintaining the Global Survival Fund website www.globalsurvivalfund.org, RFC Direct website www.rfcdirect.net, Green WALET website www.greenwalet.org, and the Risk Reconnaissance Group website www.riskreconnaissancegroup.org, and cell phone apps for each. It sells advertising on websites and apps, then collects and distributes revenue.

Advertising and sponsor revenue from events, films, websites and cell phone apps will pay for website and cell phone app administration, global event production, documentary film productions, meeting and conference expenses, and funds Green WALET and the Global Survival Fund. Website and cell phone app advertising targets billions of RFC members and Global Survival Fund donors. The accumulated advertising and sponsor revenue helps pay off RFC twenty-year bonds on their maturity dates.

Zone Interactive issues media press releases and publishes the RFC Direct and community e-newsletter that promotes the Global Survival Fund donation program and provides RFC community updates to RFC members.

RANCH FARM COOPERATIVE COMMUNITIES

Each Ranch Farm Cooperative free-trade zone green community will be a registered nonprofit foundation governed by RFC Direct and regulated for the compliance of RFC bylaws and methodology by Green WALET. The 125,416 owners of each RFC community are listed in a trust account that is established for each RFC and can be easily modified to remove or add an owner at any time to maintain the 125,416 owner-worker quota. For example, when an RFC owner resigns from their RFC and relocates outside the RFC community, they are removed as an owner on record and replaced by a new owner that assumes the previous owner's position within the RFC.

RFC Direct and Green WALET officials appoint each RFC green community foundation's Board of Directors. Each Board of Directors governs their community under the bylaws written and updated by RFC Direct and Green WALET.

Each RFC community is a democracy that elects a mayor from the residents within the community. The mayor is not required to be an RFC owner. The mayor's responsibilities include resolving any non-RFC issues and assuring those issues never interfere with RFC production and distribution. The mayor is also responsible for the community's budget for expenses that are not RFC operational expenses. The community's expenses will be paid for from an equal tax amount imposed on every resident living within the RFC green community. The mayor governs the community and oversees the police, judiciary, fire department, schools, parks and recreation, bike path maintenance, event venues, holiday entertainment, parades, street maintenance, public transportation, etc.

The mayor operates under the bylaws of the community and works with the RFC owners' officials that are appointed by the Board of Directors.

17
HEALTHCARE

Universal healthcare, aka socialized medicine, is not available for citizens in most countries. Of the 34 OECD nations, 32 have it. Mexico and the United States do not, although the U.S. has made an attempt by initiating the Affordable Care Act. Though several countries outside of the OECD have enacted some type of government-subsidized healthcare program, the majority of countries have failed to do so. [104][107][108]

Each country with universal healthcare available to their citizens has a different version. Not all coverage is the same from one country to another. Nor are accrued expenses for claims that are covered paid for in the same way. It depends on the country's program.

Regardless of how healthcare plans are implemented all countries have failed to achieve a comprehensive program. Profiting on people's ailments and injuries is big business. The prestige of being a physician with a doctor's title is highly sought after.

A doctor's medical practice is perhaps the most honorable of any profession. Healthcare is a thriving business that continues to expand as the world's population multiplies. Unfortunately, it has also become a business that has attracted the wrong people for the wrong reason. Many people that have entered the medical profession are not healers nor are they interested in becoming healers. Their sole reason is for a lucrative salary instead of a benevolent one. This voracity has been one of the main reasons universal healthcare has not happened in the U.S. Meanwhile, medical costs continue to escalate.

Healthcare reform has been a major topic for decades. Similar to discussions about climate change, water shortages, poverty, sustainable power generation, famine, etc., it has been intensely debated continually for years and everyone is aware of the issue, but a real solution for this perpetual problem has not yet been attained. There is a solution to all of these perennial topics of discussion, but resistance still continues. With healthcare it is from the medical-related organizations and associated unions that stand to lose their hold on money and power. This continues to be a stumbling block and has prevented a solution to fix the problem.

Though healthcare is an essential requirement for life, people often neglect their health. As with any mechanical device, the human body has care and maintenance requirements that need to be performed on a regular basis. If not, parts may periodically fail or wear out. There are also defective parts that may affect a person's health. A person may be born with a defective internal organ or an allergy condition that could increase their risk of health problems. For example, one heart valve artery that is undersized compared to the other three, which is genetically transferred from mother to fetus during pregnancy. A hereditary defective condition that a person is born with, which may decrease a person's longevity and increase their claims on the benefits offered by their medical insurance policy. This is considered a preexisting condition and until recently, in the United States, it was a reason for medical insurance denial. Insurance companies are now attempting to deny claims if a doctor uses a particular tool to perform a medical procedure that is not approved for use by the insurance company, restricting a doctor's medical discretion.

Both private and corporate-owned hospitals and medical centers contend with these slow payments from insurance companies, waiting for an insurance company to decide if they should or should not honor a claim. This has led to different billing for cash-paying individuals that pay at the time of the service, instead of billing an insurance company that is notoriously slow on evaluating the service that was provided prior to paying the medical invoice.

Payment for cash-paying patients may be considerably reduced compared to overinflated charges the hospital or medical center sends to the insurance company. When charging a patient's insurance company the fee escalates. That is because the insurance company takes too long to pay the charges or they question the procedure performed to the point of costing the hospital or medical center time, money and frustration to get reimbursed by the insurance company for the procedure. If the claim is denied the patient is responsible for payment.

However, hospitals and medical centers are also to blame for their non-standardized billing rates for medical procedures. Over the last several decades many hospitals and medical centers have taken advantage of their patients by charging, or should I say ridiculously overcharging, for what are considered non-complicated routine procedures. To add to this, hospitals and medical centers that are a short distance away from each other have billed insurance companies for identical procedures, but are hundreds of dollars apart in their billing charges. A simple blood test can run from $100 to over a $1000 depending on the hospital or medical center. This is another reason why medical insurance is so costly and why there is a problem in healthcare. RFC membership offers a solution to these issues.

There is only one RFC healthcare plan and it maximizes the benefits available to every member. It is a comprehensive full-service policy included in each person's membership. Normal health insurance policies, including universal healthcare policies, typically do not cover dental. Some policies will include it for an additional excessive fee. Mental health is not covered either. But what is more astonishing is the lack of preventative healthcare and health education. The lack of these two pertinent services is partly responsible for the medical profession reaping exorbitant profits from individuals later in life and the quicker deterioration of a person's health. Comprehensive services should be standardized with nothing excluded from any health insurance policy.

Preventative healthcare and health education along with dental and mental health services are included in each RFC membership. The dues will remain the same RFC standard rate regardless of a member's age or whatever preexisting ailment, injury, or genetic problem they have either developed or inherited. Comprehensive preventative healthcare and health education classes will be a mandatory requirement in all RFC community schools. See Section 18: Education.

There are four RFC hospitals and medical centers built outside of each RFC community within populated regional locations. Transportable emergency medical facilities will be parked at local fire stations; wherever there is a great distance between a hospital and a community or a small town of members. If an emergency does occur in a rural location, a RFC satellite medical center will be nearby to provide treatment. When the patient has been stabilized they will be transported to the main hospital. Emergency medical airlift drones will be used instead of traditional truck ambulances for the emergencies that are life-threatening situations. The drones will be programmed to fly directly to a hospital or a local satellite medical center. See Section 5: Houses, Buildings, Factories. [168]

18
EDUCATION

The RFC green community's education system will be structured quite differently than the other education systems throughout the world. The basic courses in every curriculum: reading, writing, mathematics, science, geography, etc., will be standardized in RFC communities. Government, history, and fine arts, music, theater, and dance will be modified to retain the cultural uniqueness of each city, region or country. Some revisions in particular subjects will be required.

One revision will be history. This is an extremely difficult subject. It is based on certain events that may be more fictionalized than actual truth, depending on the country, region, individual school district, and what is considered the truth from a historian's viewpoint, which, in most cases, is no more than an educated guess.

Unfortunately, history relies too much on written words as truth when in fact they are just one person's opinion. Even today with our modern technology – cameras, video, and audio recording capabilities – the truth is still one person's opinion compared to another. Two people may witness the same event, but have different opinions of what they witnessed when asked for their eyewitness testimony. This problem has existed throughout history. The only history that is true is the history you don't know. So history will be taught differently in RFC schools. Relying on what are known to be the proven facts, but not relying on myths that are no more than tales of fiction.

There will be four new courses added to each RFC curriculum:

1) Your hands are your brains
2) Consciousness – the consequences of one's actions
3) Health and nutrition awareness
4) Preventative security strategies

These four additional courses will be requirements that are taught throughout a child's education. The first course, "Your hands are your brains," will be taught in the early years of child development. Beginning the day a child enters the first grade, at age 6, until the end of the fourth grade, when the course is completed. It is then replaced in the fifth grade.

The second course focuses on consciousness and is taught from first grade through high school. This course will be interconnected with the third course, "Health and nutrition awareness." The two courses are interrelated because "Consciousness – the consequences of one's actions" is linked to one's own health and nutrition awareness, both mental and physical. A fundamental fourth course, "Preventative security strategies," is a terrorist and criminal activity observation and awareness course. The basics are taught in fifth grade and advanced techniques in the ninth.

RFC education program methods are different than both public and private schools. Using the combined technique of homeschooling and the traditional classroom, students will be homeschooled, but only for four hours during the day. Afterwards, students will be required to assemble with other students to develop the socializing skills that are necessary to form healthy relationships in preparation for entering into society as an adult. This reduces the classroom space needed that would normally be required, but also lets each child study in their home without distraction from other students. Each RFC will accommodate home study by installing high-speed internet and a computer with a large flat-screen monitor display in each RFC community home. The same equipment will be installed in each school's classroom within the community.

In addition, children will be tested when entering the first grade to determine which teaching methods best suit a child's aptitude. Not all children react, or learn, in the same way as other children. The majority of children adapt to traditional teaching methods, but many that are in the majority, along with those in the minority, have been found to have an improved understanding of subjects that use visual teaching methods. Regardless of the method that is used, there will be some children with an increased aptitude for understanding very intricate subjects and a lesser understanding of others. A student should not be held back because their abilities in particular subjects do not evolve as quickly as other subjects.

In the RFC school curriculum all students will be able to move into each subject at their own rate of speed. For example, a student in the fifth grade may be reading and writing at the fifth-grade level but is able to fully understand and solve complex mathematical equations at a higher-grade level. A student should not be held back in mathematics because they're in the fifth grade. Catering to each student's intellect will keep them constantly moving forward, while stimulating their minds and alleviating boredom, frustration and distraction common in classrooms. This method of combining home study with classroom study increases a child's awareness and will advance their mind further in their younger most impressionable years.

Your hands are your brains

Until children turn 10 years old their hands are their brains. I know that sentence doesn't make much sense as a standalone statement, but the person who explained this to me was very passionate about this concept. His insights into this theory helped me to understand how computers, TV, and video games became essentially today's babysitters and the unfortunate results that have left many children over the years real life impaired.

For a young child, working with their hands is a very important process that aids in the development of their brain. They help children to think and doing so prepares them to learn the skills that society expects from competent people. For some, simply changing a lightbulb can be a challenge nevertheless understanding how electricity works. That is not to say I am in any way suggesting everyone become electricians, but the basic understanding of how things work and how tools work is a benefit to each individual regardless of gender.

Children in today's world are more experienced with what is on TV, a computer, or a video game than experiencing the challenges of using their hands for something other than a remote control, cell phone, or keyboard. It is important that a child learns by using their hands to accomplish tasks that they cannot successfully accomplish without direct contact. The hands aid in the mind's development that visual awareness alone can't replicate.

Some people may argue that a simulator can give you the needed experience. That isn't the same as using your own hands to perform a task. You have to experience feeling the components and tools by touch and the only way to do that is actually doing the job. You have to experience the realities of mistakes. Those awkward times you didn't think the task all the way through, physically not just mentally. In real life there isn't always a reset button. Which brings us to the second course added to the curriculum.

Consciousness – the consequences of one's actions

Newton's Third Law of Motion states: Every action has an equal and opposite reaction. It does not apply to people. The consequences of an action can have an unintended reaction. Reactions are not predictable and can be positive or negative. Bullying is an example of an action that provokes a negative emotional response that ranges somewhere between mild and extreme. The actions of a bully can do irreparable harm to a victim. If a child is bullied severely, then the victim's reaction can be extreme. Suicide and mass murders are extreme reactions, most often involving a very troubled teen or a young adult. This desperate and violent action can be traced back to parents and an education system neglecting a child. It can be parents, classmates, teachers, counselors, or even friends that fail to notice the difficulties that a child is facing from bullying. Their failure to see the problem is one of the issues needing to be addressed, but in many cases the real tragedy is when a severe emotional problem is known and ignored. In theory, if a conscious person contemplated the action of bullying, and various consequences were foreseen, then it probably would not occur.

The potential consequences of one's own action cannot always be predicted, but using conscious reasoning can narrow it down. Children need to learn what negative and positive actions are, how to react to an action against them, and how to think forward towards the future and not in the present when contemplating an action. This is what conscious reasoning is, however, compounding the difficulty of this technique is that every action can have multiple potential outcomes.

What this essentially means is whether an action is conducted toward an individual, a situation, or a task, all possible results of a decision need to be addressed and thought out thoroughly prior to the action being initiated. Although many real-life experiences can be great educators, referring to unintended costs of mistakes, they can frequently be avoided in most cases by simply being conscious and aware. This is a subject that needs to be taught both at home from the beginning of childhood and at school. Then continually reinforced from the beginning of a child's education through young emancipated adulthood.

Parents need to take action beginning in a child's early years of development and not let this responsibility be shuffled off entirely to the school's system. Which is a system of education that was never intended to be responsible for the mental wellbeing of each individual student, but a system of learning in preparation for a future in society. Educators need to teach a method of awareness that addresses each individual personality in the class. This requires more than a yearly parent-teacher conference.

Parents and their children will both need to participate in this conscious learning process. Both will have to be instructed in awareness until the parents are able to proceed on their own in the raising of their children. This is a problem that has arisen over decades in an education system that has been unfairly thrust into raising children rather than educating them. The current education system needs to be overhauled and the damage that has been done over the last several decades needs to be repaired. The RFC community approach of updating the curriculums and standardizing their schools addresses previous problems, beginning with the parents who were neglected during their younger years by the entire system. Current public education curriculums have been designed to prepare children to join society by teaching them their ABCs, but as adults leave them unprepared to face their children's social challenges that a conscious mind can overcome on a daily basis.

Health and nutrition awareness

This is important at any age level and will be taught to children from the first grade through high school. Health and nutrition will be combined with physical education as part of the same required course.

This course shows the correlation between physical functions of the body and required nutrients – vitamins, minerals, enzymes, etc. – and their absorption during food digestion. Staying healthy is the primary focus and showing how the effects of not exercising the body, and not feeding the body the right foods to fuel it, can lead to future health issues.

As previously stated, "Consciousness – the consequences of one's actions" and "Health and nutrition awareness" are linked. It is important for children to understand that what is ingested into their body may have negative consequences immediately and later in life. The foods consumed with preservatives or large quantities of sugar can affect a child negatively and lead to obesity or hyper-activeness. Both of these conditions are less than desirable and can be avoided by conscious and informed decisions. Drugs, alcohol, cigarettes, etc., all have serious effects on an undeveloped young mind. A conscious decision can be made, but only if the child has been taught to be aware and has the knowledge to seek out the pertinent information.

These two courses have been designed to teach each child how to develop that knowledge and how to seek out and decipher information to make the conscious decisions needed to determine the consequences of their immediate actions. This prepares a child for their place in society and can prevent major lapses in judgment in the future.

Preventative security strategies

Lethal attacks on population centers can occur at any time in any public place. The people that instigate these attacks on innocent civilians are usually mentally unstable individuals, criminals, or terrorists. Whatever the motivation is for these vicious pubic attacks, the offenders are a threat to society. Violent offenses have been carried out on civilians using explosives, guns, knives, weapons of mass destruction, vehicles, etc. And the frequency of these attacks is increasing. Training students and faculty to spot suspicious activities has become necessary to assist authorities in preventing them.

The comprehensive preventative security methods that are taught include protecting people from violent attacks plus non-violent crime prevention. Identity theft is an example of a non-violent crime that can severely damage a person's credit, reputation, and livelihood. Most of the world's population is completely oblivious to what is happening in their surroundings and because of lack of training they ignore many of the signs that could prevent crimes from occurring. Vigilance training for spotting potential violent and non-violent criminal activity is used as a precaution against these types of crimes. And it can be initiated using situational awareness techniques taught in this course, which instructs students how to be aware and observe suspicious behavior then report it to the proper authorities prior to a possible criminal act proceeding.

19

ANIMAL RIGHTS AND WILDLIFE PROTECTION

I consider this one of the most important reasons for constructing these RFC communities. Unfortunately, planned community development in the past didn't really consider animal rights and migration routes in their plans. It was completely overlooked. Animals, unfortunately, do not have a voice and it is up to humans to supply that voice, although most people are completely oblivious to an animal's requirements for survival.

In the past century, because of encroachment of the human race onto wildlife habitats, animals have had to either migrate out of the area or choose to find a way to coexist with human encroachment, to stay in the regions that they and their ancestors have called home for millennia. The bears, mountain lions, alligators, coyotes, wolves, lynx, etc., that have traditionally stayed hidden from human view have been seen out in the open much more frequently and have caused problems for humans in residential developments. The negative publicity from the frequent news reports about animal attacks has neighborhood residents requesting that their local animal control enforcement authority relocate troublesome animals. However, animals are very territorial and some find their way back to the region where they had previously been removed. When this happens an animal is often killed for merely returning to its home. [165]

This policy is unacceptable and calls for better planning to avoid relocating animals from their ancestral home regions. Animals must have rights and RFCs will respect those rights. It is in the charter of every RFC green community to develop around wildlife habitats and refuge areas to avoid the conflicts that are inevitable if human encroachment continues to occur. RFCs will consider this obligation a vital stipulation during the planning stages of RFC communities and an essential guide for harmony within each green community surrounded by wildlife. [176][178]

This is non-negotiable. All wildlife, whether endangered or not, must be protected. Too many species have become extinct on this planet over the last century, mainly from human encroachment. Wildlife refuge is one of the most important aspects of every RFC community's land selection and development process and is just as important as climate change. The RRG team that is located in each RFC green community is assigned to protect endangered wildlife in addition to their other duties. See Section 16: Conglomerate. [124]

Remote-controlled solar airplanes will be used for reconnaissance in those vulnerable regions where illegal activities are conducted outside community boundaries. A solar airplane can fly for several days without landing. Those inland RFC communities that are surrounded by wildlife habitats and sanctuaries assist the local authorities in wildlife protection. The RRG team will use their observation training skills to protect the areas surrounding each RFC. Solar airplanes, flown out of RFC locations near the coast in ocean regions, will assist local authorities by providing aerial surveillance to help protect ocean wildlife from poachers and illegal fishermen on the open water, the mainland's coast, and the coastlines surrounding islands. [152]

These solar airplanes are included in the $400 billion allocated for wildlife preservation, RFC security and defense operation, and for the equipment to protect wildlife from poachers, illegal hunters and trappers, and from other human activities that threaten their existence.

20
ENVIRONMENTAL CLEANUP

For millennia the human race has made an appalling mess of our planet's environment, but all previous centuries have been nothing compared to what we have done to the environment during the last two. And today we are paying the price with the rapid acceleration of the natural progression of climate change. In short, Mother Nature has become quite irritated. And to guarantee our continued survival the human race needs to make amends. The RFC green community plan allocates considerable funding for global environmental cleanup, which includes $1 trillion for ocean cleanup, and the restoration of coral reefs, aquaculture, and wetlands.

Cleanup of coastal regions that have been affected by beaching of trash and fishing gear, such as nets, hooks, and line, must be a priority along with the elimination of the five enormous trash gyres in the Indian, North and South Atlantic, and North and South Pacific oceans. This will be tasked to fishermen whose boats will be idle after completion of the land-based fisheries. The fishing boats will be provided diesel fuel from the conversion of the plastic recovered and their exhaust will be captured for greenhouse gases; filtering out the CO_2 to combine with hydrogen to produce synthetic SNG. See Section 6: Electric Power and Fuel.

To prevent any further toxic contamination of our oceans from potential natural disasters, or from the inevitable rising water level in lowland coastal regions, it is prudent to finish cleanup of superfund sites and relocate landfills, nuclear power plants, factories using or producing toxic chemicals, etc., and stop constructing them in locations vulnerable to hurricanes and tsunamis, or below a 216-foot elevation. With the ever-increasing frequency of natural disasters striking coastal regions, and the floodwaters flowing into the oceans, this is a preventative necessity and should be a global priority. The debris from disasters is damaging enough without poisoning the water too. [5][6]

Land protection is vital for our survival and $426,108,080,000 has been allocated in this plan's global budget for cleanup of farm soil and surface water contamination, organic fertilizing, soil regeneration, and cultivation of the topsoil in the selected regions where RFC community development will occur. The cleanup includes areas in regions that will not only be within a community's boundaries, but also in areas that may affect the community in the future.

Recently the United States' Environmental Protection Agency (EPA) was at work cleaning up a superfund toxic mine site that is upriver from a Navaho Nation reservation. The contracted EPA workers made a serious error in judgment, creating an extremely hazardous situation that contaminated a major source of water on the reservation, which was a valuable river used to irrigate their farms and water livestock. This was a catastrophe of tremendous proportions and has lasting environmental consequences for the reservation and land downriver. This is a United States Government issue and provides a valuable insight for selecting locations for the RFC green communities that could be affected in the future by the inanity of the government, or other organization that has well-meaning intentions, but, if a blunder does occur, could have severe consequences. This similar situation has happened in regions throughout the world. Too many and too often to completely list here. The cautious selection of RFC land for each community must take into account what can potentially happen in the future without exception. [73]

Excessive usage and prolonged droughts caused by global climate change have reduced potable water supplies in regions and communities worldwide. To alleviate these disastrous water shortages, some desperate communities carelessly tapped into their water resources that had been polluted for decades by corporations disposing of contaminated waste. It created severe environmental damage and health problems throughout the communities. Corporations weren't concerned that the communities, which had enticed them to construct facilities in their regions to provide jobs, might need to use their water and land resources in the future. See Section 4: Water Requirements. [78][79][106][123][125]

Governments should not have to be responsible for the cleanup of contamination that was created by corporations. Unfortunately, these corporations have been allowed to contaminate land and water resources for much too long without any real consequences to the individuals that made these disastrous errors. Leaving the governments, and ultimately the taxpayers, paying the bill. If a corporation pollutes the environment they should be forced to clean it up. The executives and the stockholders of the corporation should not be allowed to hide behind the corporate bankruptcy laws to eliminate their responsibilities of cleaning up their mess. The stockholders should not be allowed to walk away from their financial loss by quickly selling their shares in the company and moving on to other investments. Investing should be treated as a shared liability and if you choose to invest in a company, you choose responsibility.

Too many times corporations have gone bankrupt and the people that were instrumental in creating the catastrophe, by making very bad decisions, should not have been allowed to walk away from their liability and keep their fortunes at the expense of the taxpayers. This is a policy that must cease globally. Stockholders that have invested in a corporation to make money should be as responsible as the directors and executives of the corporation they have demanded boost the company's earnings so the value of their investment increases. Investors would then think twice about investing in companies using technology that could prove harmful to the environment, wildlife, and the public because of the stockholders' demands placed on the company to increase its worth.

Corporations that have managed to avoid their environmental ocean and coastal cleanup responsibilities continually, by what I consider displays of courtroom antics, must be made accountable for their share of global ocean pollution. They have somehow managed to manipulate the courts into placing monetary caps on cleanup expenses. This should not be allowed. If you polluted it, you clean it up regardless of the cost, and you must continue to clean it up until it is back to its original pristine condition prior to the damage.

The costs of environmental damage should not be passed on to the consumers of any corporation's products. A mandatory requirement of price freezing on the products at the time of their verified involvement in a catastrophe must be implemented. The prices they charged their customers the day the disaster occurred until the cleanup is completed. For too long the consumer has had to foot the bill for the mistakes or, what has been proven to be more accurate, the extreme recklessness of the corporations. If a company is found guilty and fined, whether they file for bankruptcy or not, they simply raise their prices to the consumer to cover the expense. However, if corporations were prevented from this outrageous practice, then more companies would initiate major safety precautions to keep environmental damages from occurring. [79]

The environmental cleanup includes our planet's atmosphere in addition to the land and oceans. CO_2 and nitrous oxide (N_2O) pollution emitted from fossil-fueled power plants; vehicle emissions from ships, trucks, airplanes, automobiles, machinery, etc.; and small gasoline engine emissions from lawnmowers, blower motors, chain saws, generators, etc.; and the HFC and the internationally banned CFC that are emitted from worn-out refrigerators and air conditioners all contribute to greenhouse gas pollution. [131][141]

Although it has been fully banned since 2010, CFC is being used in China's home insulation industry. And it continues to damage earth's ozone layer. Even though China is a signatory on the Montreal Protocol's 1987 treaty, they are failing to enforce the ban. [161]

Not all gases emitted into the atmosphere are greenhouse gases, but still cause significant damage as smog and acid rain. These gases and particulates: N_2O, formaldehyde, non-methane organic gases, smoke, particulate matter, sulfur dioxide (SO_2), carbon monoxide (CO), etc., form smog, the brown haze reaction in sunlight harmful to air-breathing species. N_2O and SO_2 are the primary causes of acid rain, which alters the chemistry of the soil and water, affecting plants. This, however, is not the cause of ocean acidification, which is caused by CO_2 dissolving in oceans, severely affecting aquatic species. The smog that is created by fossil fuel emissions and smoke pollutes the lower atmosphere in varying degrees and is responsible for medical conditions ranging from minor breathing difficulties and watery eyes to serious respiratory infections.

Like-minded people throughout the world are actively trying to change two entire industries by eliminating the dirtiest, least efficient, most expensive, and environmentally disastrous power generation plants on the planet and the fossil fuels that power them. The power plants are used to keep these same like-minded people, and millions of others, in comfort within their homes and businesses that, until recently in the past two decades, were totally oblivious to the environmental damages power plant emissions are causing to the planet. They are also actively trying to eliminate the fossil fuels that power these plants, and the other modern conveniences that people have grown accustomed to, without a sufficient global alternative being presented. Change won't occur until an adequate solution is initiated.

Methane is much more damaging in the atmosphere than CO_2 emissions. Although it doesn't last as long, while it is there it does serious global warming damage to our planet. The three main methane emitters are cattle, followed by rice cultivation and indigenous wetlands. The RFC green communities can significantly decrease, and possibly eliminate, the emissions from cattle and rice, but global wetlands' emission cannot be reduced. When our planet's ice sheets and glaciers melt the rising ocean level will not be our greatest concern. It will be the methane and CO_2 emissions from organic matter frozen in permafrost, which accumulated over millennia, that is under the melting ice sheets and glaciers. When it is thawed, this will be our planet's number one methane-emitting source. [83][84][90][91][92][113][131][137][139]

While this happens, and it has already begun, billions of tons of methane will be released into the atmosphere. This combined with other existing greenhouse gases, which have a GWP causing us great concern, will suffocate our planet and the rise in global temperature will increase to a calamitous point. Time is of the essence. Ice cap, ice sheet and glacier meltdown are results of global warming and the only way for us to combat it, and survive, is to immediately stop the emissions of those greenhouse gases that we do control. All our emissions must cease. We need to allow enough time to rid our planet of the majority of greenhouse gases that are currently in the atmosphere. This will be decades for CH_4, N_2O, and CFC, and several centuries for CO_2, to achieve any significant reduction. Hopefully as global ice sheet meltdown continues there will be enough greenhouse gases removed from the atmosphere that increased emissions of CH_4 and CO_2 will not raise the planet's temperature to a cataclysmic degree. [8][85][86][144]

Many greenhouse gas emitters can be eliminated or converted to non-emitters. Fossil-fueled power plants can be eliminated, vehicles and small engines can be converted, and cattle burping and rice cultivation emissions can be reduced. HFC refrigerant can be replaced with NH_3 or HFO. The RFC green community plan provides solutions to these global gas emission issues.

The continued practice of burning down forests also contributes to our environmental crisis, releasing tremendous amounts of smoke and CO_2 into earth's atmosphere. Deforestation happens in heavily forested areas usually for one of three reasons: lumber, agriculture, or livestock grazing. The only way to stop this massive devastation is to remove the reasons for the deforestation. RFC green communities would remove the agriculture and livestock reasons, which eliminates the air pollution from the burning acreage for farm and ranch clearing. Lumber will continue to be an issue and the only way deforestation will stop for this purpose is for governments to place restrictions on logging. The worldwide demand for lumber cannot be eliminated, but the way logging is currently managed can be changed and the clear cutting prevented, lessening the long-term environmental impact.

If deforestation burning is stopped and emissions from the other sources cease, it doesn't signify that climate change will be terminated. Greenhouse gases will continue to be in the atmosphere. CO_2 is currently at 411 ppm. Though CH_4, N_2O, HFC, CFC, and other gases have a higher GWP than CO_2, and they don't remain in the atmosphere as long, there is a greater percentage of longer-lasting CO_2 causing significant damage to our planet. [111][134][142]

Ocean acidification is caused by excess CO_2 in the atmosphere dissolving in oceans, rivers, lakes, and manmade reservoirs, changing the pH of the water. Although our planet has immense oceans, they are still affected by the atmosphere. Since the dawn of the industrial age, the last two centuries have seen an increase in ocean acidity rise 30%, which is faster than any change in ocean chemistry in the last 50 million years.

In chemistry, the pH logarithmic scale of measurement of acidity and alkalinity in water-soluble solutions has a range of 1 thru 14 with 7 as the middle or neutral point. Values below 7 are acidic, which increases as the numbers decrease, with 1 being the most acidic. Values that are above 7 are alkaline, with 14 having the most alkalinity. Ocean pH has dropped from 8.2 to 8.1 since the industrial revolution began and is expected to go down another 0.3 to 0.4 pH value by the end of this century. Though 0.1 does not seem substantial the scale is logarithmic. A drop from pH 5 to 4 is 10 times more acidic than 5. A drop from pH 6 to 4 is 100 times more acidic. The oceans will not drop below 7, even with CO_2 dissolving at the current rate. However, a drop to 7.7 would create acidic oceans not seen for 20 million years.

Buffering is the natural pH stabilizing process of rivers carrying dissolved chemicals from rocks to the oceans, which previously kept the oceans' pH level in balance. Because of the increased levels of CO_2 in the atmosphere, and the construction of river dams, this process is no longer sufficient to keep the pH stabilized.

The result has been rapidly dropping pH in surface water, which gradually mixes with deep water and affects the entire ocean. Marine life has evolved over millions of years and the quick environmental change in the ocean ecosystem has left many species impaired to adapt to the rapid changes in water chemistry. Many chemical reactions essential for life are sensitive to small changes in pH. For example, humans have a 7.35 to 7.45 pH. A drop in blood pH of 0.2 to 0.3 could cause a seizure, coma, or death. Similarly, a small change in water pH can have cataclysmic effects on marine life impacting their chemical communication, reproduction and growth. The building of skeletons in marine creatures is particularly sensitive to the acidity. Shells of some species are already dissolving in these increasingly acidic conditions. In general, the oceans' acidification negatively impacts their ecosystems. If marine wildlife cannot adapt, then mass extinctions will occur. [181]

Our planet has a fever. When it reaches 95°F, photosynthesis, the process plants use to convert CO_2 into carbohydrates, declines quickly. Terrestrial plants will be converting far less of the greenhouse gas CO_2. Agriculture will be severely affected along with other vegetation, though this fails in comparison to the prospect of running out of breathable air. Our planet's vast oceans produce over 50% of its oxygen. The tiny marine plant phytoplankton will halt its oxygen production if ocean temperature rises up another 6°C. [2][158][171][179]

Algae cultivation is a solution to this immediate problem. Algae grow in water and can absorb considerably more CO_2 than other plants. The algae pond water can be kept below 95°F. Algae cultivation for algae oil conversion to B100 biodiesel fuel would significantly reduce the CO_2 in the atmosphere and provide enough fuel for B100 biodiesel turbines to power the planet, exceeding the current power generation rate. These are two advantages of algae cultivation and both will contribute to saving our planet from overheating. See Section 6: Electric Power and Fuel.

There are other techniques developed for capturing atmospheric CO_2 including direct air capture, manufacturing carbon nanofibers and artificial trees, reforestation, and olivine rock weathering. Once removed, combining hydrogen with the CO_2 will make synthetic SNG, or the CO_2 could be converted to diesel fuel, and olivine sand added to fertilizer. Biowaste from the cultivation of guinea grass and algae can be heated to produce biochar. These are some of the viable methods for atmospheric CO_2 removal. [65][76][114][115][116][117][118][119][163]

If we fail to immediately reduce greenhouse gases and cool our planet down, then plants will cease converting CO_2, oxygen production will decline, ocean acidification will exterminate most marine creatures, and the consequences will be devastating. Climate change is here to stay, but global warming can still be contained. Every terrestrial species now depends on human action for survival. [2][74][75][111][129][159][164][166]

21
RFC FORTY-YEAR TIMELINE

The following forty-year timeline covers what I consider the first cycle of the RFC communities. After the first forty years each community will go through a complete overhaul replacing equipment and technology to improve their efficiency. The first year begins with the twenty-year bond sale, land selection, surveying, etc. During each following year other RFC components begin, get ordered, continue, or are completed, until each RFC is fully operational at the beginning of year eight. Then begins the economic task of website advertising, commercial freight distribution, and passenger transportation, to pay off the bonds on their twenty-year maturity dates.

When nuclear fusion becomes available it will be installed at each REM site. After the installation, the conversion of gas appliances, water heaters, heating oil furnaces, etc., to electric will be mandatory for each member on a REM where increased amounts of electricity are generated.

YEAR 1

Begin:
Twenty-year bond sale
Land selection
      Grant of free-trade zone status for land selected
      Purchase land for RFC green communities
      Purchase land for Maglev transit route, water pipeline route
      Purchase land for Maglev distribution terminals and REM sites
Survey of land, transit and pipeline route, distribution terminals, REM
      Site plans for each RFC community
      Architect and engineering-approved building plans
Owner-worker training
      Construction and factory
      Farming and agro-ecology
      Rancher livestock selection and breeding
Outside vendor tool up for RFC construction

Underground transit tunnel construction survey
> Bering Strait (already begun)
> Strait of Gibraltar
> Bab-el-Mandeb

Concrete plant construction
Selecting computer programmers

Order:
Power-generation accessories
> Anaerobic digesters
> Fermentation equipment
> Power turbines

Ocean-desalination plants
Autoclaves and mixers for cattle feed
Airships 500T capacity
Sewage treatment plants

YEAR 2

Complete:
Grant of free-trade zone status for land selected
Purchase land for RFC green communities
Purchase land for Maglev transit route, water pipeline route
Purchase land for Maglev distribution terminals and REM sites

Continue:
Twenty-year bond sale
Survey land, transit route, pipeline route, distribution terminals, REM
> Site plans for each RFC community
> Architect and engineering-approved building plans

Plant fruit trees
Plant moringa and guinea grass
Outside vendor tool up for RFC and Maglev transit construction
Owner-worker training
> Construction and factory
> Farming and agro-ecology
> Rancher livestock selection and breeding

Concrete plant construction

150

Begin:
Installation for power plants
Installation for ocean-desalination plants and pipeline
Selection of livestock
Underground transit tunnel construction
> Bering Strait (already begun)
> Strait of Gibraltar
> Bab-el-Mandeb
Computer programming

YEAR 3

Complete:
Survey of land, transit route, pipeline route, distribution terminals, REM
> Site plans for each RFC community
> Architect and engineering-approved building plans
Continue:
Twenty-year bond sale
Underground transit tunnel construction
Installation for power plants
Installation for ocean-desalination plants and pipeline
Livestock selection
Owner-worker training
> Construction and factory
> Farming and agro-ecology
> Rancher livestock selection and breeding
Begin:
RFC owner-worker selection process for RFC community owners
Owner-worker house construction
Freshwater and saltwater fishery construction
Land cleanup for agriculture, housing and community development
Hospital and school construction
Construction of RFC livestock ranches
Agriculture for livestock feed
Construction of livestock feed processing plant
Livestock breeding

Maglev transit and distribution terminal construction
      Flatcar and container manufacturing
Datacenter construction
REM sites and communication network
Order:
Fuel cells for datacenter

YEAR 4

Continue:
Twenty-year bond sale
Owner-worker house construction
Hospital and school construction
Freshwater and saltwater fishery construction
Land cleanup for agriculture, housing and community development
Underground transit tunnel construction
Maglev transit and distribution terminal construction
      Flatcar and container manufacturing
Installation for power plants
Installation for ocean-desalination plants and pipeline
Livestock selection
Construction of RFC livestock ranches
Agriculture for livestock feed
Construction of livestock feed processing plant
Livestock breeding
Owner-worker training
      Construction and factory
      Farming and agro-ecology
      Rancher livestock selection and breeding
Computer programming
Datacenter construction
REM sites and communication network
Begin:
Community-building construction
Factory-building and intra-RFC distribution construction
Crop selection and planting for agriculture to boost nutrients in soil

Order:
Factory equipment for installation in fifth year

YEAR 5

Continue:
Twenty-year bond sale
Owner-worker house construction
Hospital and school construction
Freshwater and saltwater fishery construction
Land cleanup for agriculture, housing and community development
Underground transit tunnel construction
Maglev transit and distribution terminal construction
       Flatcar and container manufacturing
Installation for power plants
Installation for ocean-desalination plants and pipeline
Livestock selection
Construction of RFC livestock ranches
Agriculture for livestock feed
Construction of livestock feed processing plant
Livestock breeding
Owner-worker training
       Construction and factory
       Farming and agro-ecology
       Rancher livestock selection and breeding
Community-building construction
Factory-building and intra-RFC distribution construction
Crop selection and planting for agriculture to boost nutrients in soil
Computer programming
Datacenter construction
REM sites and communication network
Begin:
Factory equipment installation
Begin crop rotation procedures
Ocean cleanup

YEAR 6

Continue:
Twenty-year bond sale
Owner-worker house construction
Hospital and school construction
Freshwater and saltwater fishery construction
Land cleanup for agriculture, housing and community development
Underground transit tunnel construction
Maglev transit and distribution terminal construction
   Flatcar and container manufacturing
Installation for power plants
Installation for ocean-desalination plants and pipeline
Livestock selection
Construction of RFC livestock ranches
Agriculture for livestock feed
Construction of livestock feed processing plant
Livestock breeding
Owner-worker training
   Construction and factory
   Farming and agro-ecology
   Rancher livestock selection and breeding
Community-building construction
Factory-building and intra-RFC distribution construction
Factory equipment installation
Crop selection and planting for agriculture to boost nutrients in soil
Crop rotation procedures
Computer programming
Datacenter construction
REM sites and communication network
Ocean cleanup

YEAR 7

Complete:
Twenty-year bond sale
Owner-worker house construction

Hospital and school construction
Freshwater and saltwater fishery construction
Land cleanup for agriculture, housing and community development
Underground transit tunnel construction
      Strait of Gibraltar
      Bab-el-Mandeb
Maglev transit and distribution terminal construction
      Flatcar and container manufacturing
Installation for power plants
Installation for ocean-desalination plants and pipeline
Construction of RFC livestock ranches
Construction of livestock feed processing plant
Community-building construction
Factory-building and intra-RFC distribution construction
Factory equipment installation
Computer programming
Datacenter construction
REM sites and communication network
Continue:
Livestock selection
Agriculture for livestock feed
Livestock breeding
Owner-worker training
      Construction and factory
      Farming and agro-ecology
      Rancher livestock selection and breeding
Crop selection and planting for agriculture to boost nutrients in soil
Crop rotation procedures
Underground transit tunnel construction
      Bering Strait
Ocean cleanup
Begin:
RFC member sign up
First year crop planting for RFC owner-workers
Factory equipment operational training for RFC owners-workers
Operation of ocean-desalination plants and pipeline
Operation of power plants

YEAR 8

Continue:
Underground transit tunnel construction
      Bering Strait
Operation of ocean-desalination plants
Operation of power plants
Ocean cleanup
Begin:
Operation of RFC farm and ranch production
Scheduled operation of Maglev transit network
Datacenter
REM sites and communication network

YEARS 9 - 11

Complete:
Underground transit tunnel construction
      Bering Strait
Continue:
Operation of RFC
Operation of Maglev transit network
Operation of ocean-desalination plants
Ocean cleanup

YEARS 12 - 19

Complete:
Ocean cleanup
Continue:
Operation of RFC
Operation of Maglev transit network
Operation of ocean-desalination plants

YEAR 20

Continue:
Operation of RFC
Operation of Maglev transit network
Operation of ocean-desalination plants
Begin:
Payout of RFC twenty-year bonds as they become due

YEARS 21 - 26

Continue:
Operation of RFC
Operation of Maglev transit network
Operation of ocean-desalination plants
Payout of RFC twenty-year bonds as they become due

YEAR 27

Complete:
Payout of final RFC twenty-year bonds as they become due
Continue:
Operation of RFC
Operation of Maglev transit network
Operation of ocean-desalination plants

YEARS 28 - 34

Continue:
Operation of RFC
Operation of Maglev transit network
Operation of ocean-desalination plants
Begin:
Maglev and advertising revenue split
      Add to future RFC equipment replacement fund
      Fund college education
      Fund mosquito eradication and medical technology
      Fund space and off-world R&D, and nuclear waste removal

YEARS 35 - 40

Continue:
Operation of RFC
Operation of Maglev transit network
Operation of ocean-desalination plants
Maglev and advertising revenue split
      Add to future RFC equipment replacement fund
      Fund college education
      Fund mosquito eradication and medical technology
      Fund space and off-world R&D, and nuclear waste removal
Begin replacing or upgrading:
RFC factory, agriculture, and livestock equipment
Power plant turbines and ocean-desalination plant equipment
Maglev transit, flatcars, and equipment

AMERICAN-TO-METRIC CONVERSIONS

LINEAR CONVERSION

feet to meters
1 ft. x 0.3048 = 0.3048 m
yards to meters
1 yd. x 0.9144 = 0.9144 m
miles to kilometers
1 mile x 1.609344 = 1.609344 km

SQUARE CONVERSION

square feet to square meters
1 sq. ft. x 0.09290304 = 0.09290304 sq. m
square yards to square meters
1 sq. yd. x 0.8361274 = 0.8361274 sq. m
square miles to square kilometers
1 sq. mile x 2.589988 = 2.589988 sq. km
acres to hectares
1 acre x 0.4046856 = 0.4046856 hectare

CUBIC CONVERSION

cubic feet to cubic meters
1 cu. ft. x 0.02831685 = 0.02831685 cu. m
cubic yards to cubic meters
1 cu. yd. x .7645549 = .7645549 cu. m

VOLUME CONVERSION

gallons to liters
1 gallon x 3.785412 = 3.785412 liters

WEIGHT CONVERSION

pounds to kilograms
1 lb. x 0.4535924 = 0.4535924 kg
tons to metric tons
1 ton x .9071847 = .9071847 metric ton
pounds per square inch to kilopascals
1 psi x 6.89475729 = 6.89475729 kPa

ABBREVIATIONS, ACRONYMS, AND SYMBOLS

AC	Alternating Current
A.I.	Artificial Intelligence
aka	Also known as
app	Application for cell phone and computer
ATV	All-Terrain Vehicle
B100	Biodiesel designation
BTU	British Thermal Unit
C	Celsius or Centigrade temperature scale
CAES	Compressed Air Energy Storage
CBM	Compressed Biomethane
CCS	Carbon Capture and Storage
CFC	Chlorofluorocarbon
CH_4	Methane
CHP	Combined Heat and Power
CNG	Compressed Natural Gas
CO	Carbon Monoxide
CO_2	Carbon Dioxide
CPV	Concentrating Photovoltaic
CSP	Concentrating Solar Power
cu.	Cubic
DC	Direct Current
EMT	Emergency Medical Technician
E100	Ethanol designation
EPA	Environmental Protection Agency
ESS	Energy Storage Solution
E.U.	European Union
F	Fahrenheit temperature scale
ft.	Foot or feet
GGE	Gasoline Gallon Equivalent
GMO	Genetically Modified Organism
GW	Gigawatt of electricity. One billion watts of electricity.
GWh	Gigawatt hour of electricity.
GWP	Global Warming Potential
HCPVT	High-Concentration Photovoltaic Thermal

HFC	Hydrofluorocarbon
HFO	Hydrofluoroolefin
hr.	Hour
H^2O	Water
JCI	Joint Commission International
km	Kilometer. One thousand meters.
kW	Kilowatt of electricity. One thousand watts of electricity.
kWh	Kilowatt hour of electricity.
LAES	Liquid Air Energy Storage
lb.	Pound
LBM	Liquid Biomethane
Li	Chemical element of Periodic Table symbol for lithium.
LNG	Liquid Natural Gas
LPG	Liquefied Petroleum Gas aka Propane
m	Meter
Maglev	Magnetic levitation
mph	Miles per hour
MSES	Molten Salt Energy Storage
MW	Megawatt of electricity. One million watts of electricity.
MWh	Megawatt hour of electricity.
Na	Chemical element of Periodic Table symbol for sodium.
NASA	National Aeronautics and Space Administration
NEO	Near-Earth Object
NGO	Non-Governmental Organization
NH_3	Ammonia
Ni	Chemical element of Periodic Table symbol for nickel.
N_2O	Nitrous Oxide
OECD	Organization for Economic Cooperation and Development
OTEC	Ocean Thermal Energy Conversion
OWC	Oscillating Water Column
PDCO	Planetary Defense Coordination Office
pH	Potential of Hydrogen
PHS	Pumped Hydro Storage
ppm	Parts per million
psi	Pounds per square inch
PV	Photovoltaic

R&D	Research and Development
REM	RFC Electric Micro-grid
RFC	Ranch Farm Cooperative
RRG	Risk Reconnaissance Group
smog	<u>Sm</u>okey <u>Fo</u>g. Brown haze air.
SMR	Small Modular Nuclear Reactor
SNG	Substitute Natural Gas
SO₂	Sulfur Dioxide
sq.	Square
T	Ton. Two thousand pounds.
TCLP	Toxicity Characteristics Leach Procedure
U.N.	United Nations
U.S.	United States
USD	United States Denomination
USDA	United States Department of Agriculture
V	Chemical element of Periodic Table symbol for vanadium.
W	Watt of electricity.
Wh	Watt hour of electricity.
WIPP	Waste Isolation Pilot Plant
WMD	Weapons of Mass Destruction
yd.	Yard
3D	Three-Dimensional
°	Earth coordinate symbol for degree. 0° Latitude
°	Temperature scale symbol for degree. 1° Celsius
$	United States Dollar currency symbol. $1
€	European Union Euro currency symbol. €1
%	Percent
[0]	Reference number

REFERENCES

[1] http://unfccc.int/paris_agreement/items/9444.php

[2] http://www.resilience.org/stories/2015-12-10/what-worries-the-world-s-most-famous-climate-scientist#

[3] http://www.usdebtclock.org/index.html#

[4] https://www.cia.gov/library/publications/the-world-factbook/fields/2079.html

[5] http://www.nytimes.com/2016/02/23/science/sea-level-rise-global-warming-climate-change.html?_r=0

[6] http://www.mymodernmet.com/profiles/blogs/national-geographic-rising-seas

[7] https://peakoil.com/enviroment/worlds-largest-fusion-reactor-will-harness-the-power-of-the-sun

[8] http://www.ipsnews.net/2011/09/200-million-depend-on-melting-glaciers-for-water/

[9] http://articles.mercola.com/sites/articles/archive/2015/08/24/moringa-tree-uses.aspx

[10] http://wholegrainscouncil.org/whole-grains-101

[11] http://markets.businessinsider.com/news/stocks/high-tech-algae-farming-industry-gets-boost-with-introduction-of-bipartisan-algae-agriculture-act-of-2018-1019194144

[12] http://www.nrel.gov/docs/legosti/fy98/24190.pdf

[13] http://www.lcbamarketing.com/biodiesel_from_algae.htm

[14] http://www.foodandwaterwatch.org/insight/factory-fish-farming

[15] https://www.livescience.com/52965-groundwater-resources-map.html

[16] http://www.takepart.com/article/2015/12/10/plastic-bottle-homes

[17] http://www.bumrungrad.com

[18] http://suscon.org/cowpower/biomethaneSourcebook/Chapter_4.pdf

[19] http://www.backwoodshome.com/articles2/yago103.html

[20] http://www.gizmag.com/hydrogen-plant-waste/36903/

[21] http://www.gizmag.com/hydrogen-production-methane-without-co2/40502/

[22] http://www.renewableenergyworld.com/articles/2014/07/hydrogen-energy-storage-a-new-solution-to-the-renewable-energy-intermittency-problem.html

[23] https://en.wikipedia.org/wiki/List_of_countries_by_electricity_production

[24] http://www.bloomenergy.com/fuel-cell/es-5710-data-sheet/

[25] https://en.wikipedia.org/wiki/Fuel_cell

[26] http://energy.gov/eere/fuelcells/types-fuel-cells

[27] https://www1.eere.energy.gov/solar/pdfs/solar_timeline.pdf

[28] http://inventors.about.com/od/timelines/a/Photovoltaics_2.htm

[29] http://news.investors.com/IBD-Editorials-Perspective/120815-784340-wind-and-solar-cannot-replace-traditional-energy-sources.htm

[30] http://cleantechnica.com/2015/05/29/alaskan-energy-storage-project-will-use-flywheels/

[31] http://www.akenergyauthority.org/Content/Programs/EETF/Documents/Round_2/Multi-Stage%20Flywheel-Battery%20Energy%20Storage%20System.pdf

[32] http://www.theoildrum.com/node/8428

[33] http://www.resilience.org/stories/2015-11-30/for-storing-electricity-utilities-are-turning-to-pumped-hydro

[34] http://energystorage.org/compressed-air-energy-storage-caes

[35] https://www.netl.doe.gov/File%20library/research/coal/energy%20systems/gasification/Proj438.pdf

[36] http://www.world-nuclear.org/information-library/safety-and-security/safety-of-plants/fukushima-accident.aspx

[37] http://www.latimes.com/opinion/editorials/la-ed-1125-san-onofre-20151130-story.html

[38] http://nuclearradiophobia.blogspot.com/p/cost-of-nuclear-power.html

[39] http://www.thebreakthrough.org/index.php/programs/energy-and-climate/nuclear-costs-in-context

[40] http://www.techtimes.com/articles/3412/20140213/giant-laser-brings-nuclear-fusion-power-one-step-closer-to-reality.htm

[41] https://www.iter.org/

[42] http://www.internationalrivers.org/environmental-impacts-of-dams

[43] http://www.theverge.com/2015/5/16/8615089/vortex-bladeless-wind-turbines-shake-to-generate-electricity

[44] http://www.huffingtonpost.com/x-prize-foundation/wiggling-wind-power_b_7888602.html

[45] http://www.vortexbladeless.com/technology.php

[46] http://news.nationalgeographic.com/news/energy/2014/11/141111-solar-panel-manufacturing-sustainability-ranking/

[47] http://spectrum.ieee.org/green-tech/solar/solar-energy-isnt-always-as-green-as-you-think

[48] http://www.worldwatch.org/node/5650

[49] http://www.technologyreview.com/view/513986/ibm-solar-dish-does-double-duty/

[50] http://www.computerworld.com/article/2687236/ibms-solar-concentrator-can-produce-energy-clean-water-and-ac.html

[51] https://phys.org/news/2012-03-thermosolar-power-station-spain-night.html

[52] http://reneweconomy.com.au/2013/salt-based-solar-thermal-plant-takes-shape-in-nevada-28157

[53] http://www.rnp.org/node/wave-tidal-energy-technology

[54] http://www.huffingtonpost.com/x-prize-foundation/innovative-seaside-power_b_8764478.html

[55] http://www.makai.com/

[56] http://www.huffingtonpost.com/ragnheiaur-elan-arnadattir/the-little-icelandic-example-pursuing-a-clean-energy-agenda_b_8771780.html?ncid=txtlnkusaolp00000592

[57] http://geothermalresourcescouncil.blogspot.com/2015/11/science-technology-using-super-critical.html

[58] http://www.truthdig.com/report/item/dragon_water_could_power_the_planet_20151106

[59] http://journeytoforever.org/ethanol_energy.html

[60] http://biodiesel.org/what-is-biodiesel/biodiesel-basics

[61] http://www.pbs.org/wgbh/nova/next/tech/beccs/

[62] http://www.businessinsider.com/a-dutch-city-is-planning-to-build-roads-from-recycled-plastic-2015-7

[63] http://www.solazymeindustrials.com/

[64] http://www.technologyreview.com/news/409654/fuel-from-algae/

[65] http://news.stanford.edu/news/2013/february/reducing-carbon-dioxide-021513.html

[66] http://www.nrel.gov/docs/legosti/fy98/24190.pdf

[67] http://www.zdnet.com/article/bye-bye-imported-oil-new-tech-converts-junk-plastics-into-fuel/

[68] http://www.afdc.energy.gov/fuels/fuel_comparison_chart.pdf

[69] https://www.capstoneturbine.com

[70] http://www.bloomberg.com/news/articles/2015-12-03/electric-cars-can-t-take-the-cold

[71] http://www.gizmag.com/shipping-pollution/11526/

[72] http://thinkprogress.org/climate/2015/12/10/3729928/bill-gates-paris-climate-research-fund/

[73] https://www.rt.com/usa/312499-toxic-spill-tribes-epa/

[74] http://finance.yahoo.com/news/nasa-satellites-reveal-something-startling-171000679.html

[75] http://www.takepart.com/article/2015/12/03/climate-change-oxygen-ocean

[76] http://www.sciencealert.com/scientists-are-building-a-system-that-could-turn-atmospheric-co2-into-fuel

[77] http://www.wired.co.uk/news/archive/2010-09/20/into-eternity-nuclear-waste-finland

[78] http://www.huffingtonpost.com/entry/flint-water-crisis-national-impact_us_56a14b47e4b076aadcc5cc52?

[79] https://www.publicintegrity.org/2017/08/17/21086/industrial-waste-pollutes-america-s-drinking-water

[80] http://www.theguardian.com/science/2005/dec/07/spaceexploration.research?CMP=share_btn_link

[81] http://www.foxnews.com/science/2018/06/20/nasa-unveils-new-plan-to-protect-earth-from-asteroids.html

[82] http://www2.mazda.com/en/technology/env/hre/

[83] https://www.aip.org/history/climate/floods.htm

[84] https://www.skepticalscience.com/methane-and-global-warming.htm

[85] http://www3.epa.gov/climatechange/ghgemissions/gases.html

[86] http://blogs.edf.org/climate411/2008/02/26/ghg_lifetimes/

[87] https://whatisnuclear.com/articles/thorium.html

[88] http://www.world-nuclear.org/information-library/current-and-future-generation/thorium.aspx

[89] http://large.stanford.edu/courses/2013/ph241/micks2/

[90] http://science.time.com/2013/07/24/antarctica-melted-in-the-past-and-as-the-climate-warms-its-poised-to-melt-again/

[91] http://newsletters.britannica.com/venda/The%20Environment_retreatingice_spread.pdf

[92] https://en.wikipedia.org/wiki/Arctic_methane_release

[93] http://www.dispatch.com/content/stories/insight/2016/03/27/01-climate-change-study-suggests-earth-is-heading-toward-a-second-catastrophic-hot-house-event.html

[94] http://www.scientificamerican.com/article/global-warming-hiatus-debate-flares-up-again/

[95] http://www.investors.com/politics/editorials/another-climate-alarmist-admits-real-motive-behind-warming-scare/

[96] http://theconversation.com/drought-proofing-perth-the-long-view-of-western-australian-water-36349

[97] http://www.pmmhf.com/article/show_article.php?id=364

[98] https://www.epa.gov/chp/what-chp

[99] http://www.worldwatch.org/node/5924

[100] http://www.businessinsider.com/europe-renewable-energy-grid-problem-2016-3

[101] https://www.duke-energy.com/about-energy/generating-electricity/nuclear-how.asp

[102] http://www.csmonitor.com/Science/2016/0329/Huge-comet-unexpectedly-whizzes-past-Earth

[103] https://www.cnet.com/news/asteroid-meteorite-impact-study-space-san-francisco-chicago/

[104] http://www.theatlantic.com/international/archive/2012/06/heres-a-map-of-the-countries-that-provide-universal-health-care-americas-still-not-on-it/259153/

[105] http://finance.yahoo.com/news/worlds-deadliest-animal-isnt-shark-161002207.html

[106] http://www.eoearth.org/view/article/153161/

[107] http://apecsec.org/socialized-medicine-pros-and-cons/

[108] http://www.oecd.org/about/membersandpartners/list-oecd-member-countries.htm

[109] https://www.oregon.gov/osp/SFM/docs/CR2K/GasConversionChartPublic.pdf

[110] https://www.usnews.com/news/best-states/articles/2018-02-16/as-world-eyes-cape-town-water-crisis-texas-study-explores-new-options?int=news-rec

[111] https://www.yahoo.com/finance/news/earths-atmosphere-traveling-back-time-002100277.html

[112] https://www.washingtonpost.com/news/energy-environment/wp/2016/06/20/a-huge-science-debate-is-brewing-over-whether-weve-messed-up-the-atlantic-oceans-circulation/

[113] https://nsidc.org/cryosphere/frozenground/methane.html

[114] http://planetsave.com/2014/04/21/technologies-direct-removal-atmospheric-carbon/

[115] https://www.theguardian.com/sustainable-business/2015/jul/14/carbon-direct-air-capture-startups-tech-climate

[116] https://www.technologyreview.com/s/540706/researcher-demonstrates-how-to-suck-carbon-from-the-air-make-stuff-from-it/

[117] https://news.vice.com/article/this-material-can-remove-carbon-pollution-from-the-atmosphere

[118] http://america.aljazeera.com/articles/2014/4/12/can-carbon-dioxideremovalsavetheplanet.html

[119] http://renergy.com/carbon-dioxide-removal-4-methods/

[120] http://phys.org/news/2015-02-negative-carbon-dioxide-atmosphere.html

[121] http://articles.mercola.com/sites/articles/archive/2014/01/12/grazing-cows-biological-farming.aspx

[122] http://www.tsp-data-portal.org/Breakdown-of-Electricity-Generation-by-Energy-Source#tspQvChart

[123] https://www.washingtonpost.com/national/health-science/us-megadroughts-are-likely-to-become-more-common-study-concludes/2016/10/07/8ded26b6-8ca6-11e6-bff0-d53f592f176e_story.html

[124] http://sportsday.dallasnews.com/other-sports/outdoors/2016/10/08/urban-bobcats-roam-secretly-avoid-humans-shy-pets

[125] http://www.sfgate.com/news/article/SF-Bay-ecosystem-collapsing-as-rivers-diverted-9953776.php?cmpid=twitter-desktop

[126] http://www.pbs.org/newshour/rundown/oregon-transform-lake-battery-charge-electricity-grid/

[127] https://www.newscientist.com/article/2108296-first-farm-to-grow-veg-in-a-desert-using-only-sun-and-seawater/

[128] http://www.theatlantic.com/science/archive/2016/10/megadroughts-arizona-new-mexico/503531/?utm_source=yahoo&yptr=yahoo

[129] http://www.businessinsider.com/climate-change-plays-out-before-us-2016-8

[130] http://www.space.com/33891-newfound-asteroid-buzzes-earth-2016-qa2.html

[131] http://www.nytimes.com/2015/09/12/science/climate-study-predicts-huge-sea-level-rise-if-all-fossil-fuels-are-burned.html?action=click&contentCollection=Science&module=Related Coverage&region=EndOfArticle&pgtype=article

[132] http://forms.madisonlogic.com/Form.aspx?pub=815&pgr=1472&frm=2278&autodn=1&src=15273&ctg=667&ast=58418&crv=0&cmp=17069&yld=0&clk=6291703552715132308

[133] https://www.linkedin.com/pulse/microgridding-us-one-state-time-elisa-wood?trk=prof-pos

[134] http://web.unep.org/ozonaction/who-we-are/about-montreal-protocol

[135] http://www.cbsnews.com/news/obama-calls-paris-agreement-a-turning-point-for-our-planet/

[136] http://www.marketwatch.com/story/22000-years-of-climate-change-captured-by-a-cartoonist-2016-09-13?siteid=yhoof2&yptr=yahoo

[137] http://www.cnn.com/2016/09/20/opinions/sutter-arctic-melting-george-divoky/

[138] https://www.wired.com/2016/10/forget-hyperloop-awful-looking-thing-can-move-stuff/

[139] https://www.washingtonpost.com/news/energy-environment/wp/2016/10/11/how-greenlands-ice-is-melting-from-above-and-below/?utm_term=.75c39bb3dc04

[140] http://www.zdnet.com/article/bye-bye-batteries-mits-new-power-sponge-could-hold-key-to-electric-cars/

[141] http://www.greenpeace.org/international/Global/international/planet-2/report/2009/5/HFCs-Fgases.pdf

[142] http://cdiac.ornl.gov/pns/current_ghg.html

[143] http://www.livescience.com/38519-refrigerant-hfcs-devastating-to-climate.html

[144] https://www.theguardian.com/environment/2012/jan/16/greenhouse-gases-remain-air

[145] https://climate.emerson.com/documents/presentations-chicago-%E2%80%93-natural-refrigerant-alternatives-for-industrial-refrigeration-en-us-3663318.pdf

[146] http://www.globalissues.org/news/2016/11/11/22637

[147] http://www.wipp.energy.gov

[148] https://www.forbes.com/sites/mikescott/2018/06/08/liquid-air-technology-offers-prospect-of-storing-energy-for-the-long-term/#5b73d9ba622f

[149] https://www.npr.org/2018/06/06/617532884/aisle-middle-or-video-emirates-president-predicts-windowless-planes-are-coming

[150] https://www.forbes.com/sites/jamesconca/2018/05/15/nuscales-small-modular-nuclear-reactor-passes-biggest-hurdle-yet/#4a3b96885bb5

[151] http://batteryuniversity.com/learn/article/bu_210b_flow_battery

[152] https://www.engadget.com/2018/08/13/airbus-solar-powered-zephyr-drone-record/?yptr=yahoo

[153] https://www.aljazeera.com/indepth/features/2017/01/child-policy-affected-china-170129130503972.html

[154] http://www.bbc.com/earth/story/20170504-there-are-diseases-hidden-in-ice-and-they-are-waking-up

[155] https://techcrunch.com/2018/10/03/iron-ox-opens-its-first-fully-autonomous-farm/?yptr=yahoo

[156] https://www.iris.xyz/learn/global/will-global-market-crash-2018

[157] https://spacenews.com/advisory-group-skeptical-of-nasa-lunar-exploration-plans/

[158] http://www.takepart.com/article/2015/12/03/climate-change-oxygen-ocean/

[159] https://www.globalchange.gov/

[160] https://www.engadget.com/2018/10/21/electric-car-carbon-fiber-battery/?yptr=yahoo

[161] https://www.bbc.com/news/science-environment-44738952

[162] https://www.airships.net/helium-hydrogen-airships/

[163] https://www.businessinsider.com/r-sucking-carbon-from-air-swiss-firm-wins-new-funds-for-climate-fix-2018-8?amp%253Butm_medium=referral

[164] https://www.mercurynews.com/2018/08/10/letter-time-is-not-on-our-side-to-act-on-climate-change/

[165] http://www.nhptv.org/natureworks/nwep12a.htm

[166] https://www.businessinsider.com/global-warming-point-of-no-return-temperature-2018-8?amp%253Butm_medium=referral

[167] https://www.washingtonpost.com/world/europe/un-says-more-worldwide-going-hungry-blames-climate-change/2018/09/11/fd521652-b5b1-11e8-ae4f-2c1439c96d79_story.html

[168] http://alothome.com/landing?slk=drone+ambulance&nid=2&cid=79439735022584&kwid=79439743413876&akwd=%2Bdrone%20%2Bambulance&dmt=e&bmt=bb&dist=s&uq=Drone%20ambulance&device=c&ismobile=false&msclkid=4ef13c2a64431875da0c676980f82aac&accid=35000746&campid=291303583&agid=1271035509027386&vx=0

[169] http://www.foxnews.com/science/2018/05/18/stunning-nasa-study-shows-humans-are-responsible-for-major-changes-to-earths-water-availability.html?utm_source=feedburner&utm_medium=feed&utm_campaign=Feed%3A+foxnews%2Fscitech+%28Internal+-+SciTech+-+Mixed%29

[170] https://www.yahoo.com/finance/video/robots-changing-farming-industry-123700434.html

[171] https://mashable.com/2018/05/03/co2-highest-level-human-history/#MOoUyUemXsqi

[172] https://www.cnn.com/travel/article/dubai-vertical-farm-emirates-catering/index.html?utm_source=feedburner&utm_medium=feed&utm_campaign=Feed%3A+rss%2Fcnn_latest+%28RSS%3A+CNN+-+Most+Recent%29

[173] https://www.sfgate.com/news/science/article/Study-Global-warming-is-weakening-key-ocean-12825506.php

[174] https://www.whitehouse.gov/wp-content/uploads/2018/06/National-Near-Earth-Object-Preparedness-Strategy-and-Action-Plan-23-pages-1MB.pdf

[175] http://markets.businessinsider.com/news/stocks/high-tech-algae-farming-industry-gets-boost-with-introduction-of-bipartisan-algae-agriculture-act-of-2018-1019194144

[176] https://www.yahoo.com/news/disturbing-composite-photos-reimagine-ocean-slideshow-wp-184651080.html

[177] http://www.foxnews.com/great-outdoors/2018/01/24/fisherman-makes-horrific-find-in-stomach-mahi-mahi.html

[178] https://www.forbes.com/sites/trevornace/2018/04/09/yet-another-dead-whale-is-grave-reminder-of-our-massive-plastic-problem/#6082d2196cd2

[179] https://www.upi.com/Science_News/2018/01/04/Oxygen-levels-in-Earths-oceans-continue-to-drop/8031515097899/?st_rec=5071519306116

[180] https://mashable.com/2018/08/18/how-recycling-works/

[181] https://ocean.si.edu/ocean-life/invertebrates/ocean-acidification

[182] http://www.bccrwe.com/index.php/8-news/9-are-wind-turbine-rare-earth-minerals-too-costly-for-environment